Synthesis Lectures on Technology Management & Entrepreneurship

In the dynamic intersection of technology innovation and entrepreneurship, organizations navigate a rapidly evolving landscape. Books in the series **Synthesis Lectures on Technology Management & Entrepreneurship** are aimed at helping readers with different backgrounds understand topics in the field. Coverage includes successful entrepreneurship, effective management of emerging technologies to deal with uncertainty and capitalize on opportunities, management of innovation processes, and approaches to improve performance and foster a climate of continuous innovation and adaptation in the organization.

Varun Gupta • Chetna Gupta

Technological Innovations in Education

Generative Artificial Intelligence (GenAI) in Academia

Springer

Varun Gupta
Multidisciplinary Research Centre for Innovations in SMEs (MrciS)
Gisma University of Applied Sciences
Potsdam, Germany

Department of Economics and Business Administration
University of Alcala, Alcalá de Henares (Madrid)
Madrid, Spain

Chetna Gupta
School of Computer Science Engineering and Technology
Bennett University
Noida, Uttar Pradesh, India

ISSN 1933-978X ISSN 1933-9798 (electronic)
Synthesis Lectures on Technology Management & Entrepreneurship
ISBN 978-3-032-11224-8 ISBN 978-3-032-11225-5 (eBook)
https://doi.org/10.1007/978-3-032-11225-5

This Springer imprint is published by the registered company Springer Nature Switzerland AG
The registered company address is: Gewerbestrasse 11, 6330 Cham, Switzerland

My book is dedicated to my parents as well as the almighty Goddesses Koyla, Myah Bhagwati, Jalpa Mata, and Naina Mata.
Mi libro está dedicada a mis padres y a las diosas todopoderosas Koyla, Myah Bhagwati, Jalpa Mata, y Naina Mata.
Mein Buch ist sowohl meinen Eltern als auch den allmächtigen Göttinnen Koyla, Myah Bhagwati, Jalpa Mata, und Naina Mata gewidmet.

Preface

The rapid rise of generative artificial intelligence (GenAI) is reshaping education in ways that could not have been imagined just a few years ago. For universities, the challenge is not whether students will use GenAI, but how it can be meaningfully and ethically integrated into the courses. This book was written out of that challenge and out of real practical experience of integrating the GenAI in postgraduate courses, monitoring its outcomes, and refining its role as a "knowledge collaborator" with non-substantial contribution rather than a substitute for student effort.

The research work began with a simple question: How can students learn to engage with GenAI responsibly while still meeting their intended learning outcomes and upholding academic integrity? The answer required a redesign of courses, assignments, and evaluation practices, supported by clear AI course policies, GenAI usage instructions, and practical instruments like GenAI integration rubrics. Over multiple semesters, students were introduced to structured experimentation with GenAI and guided to reflect on its strengths and limitations through developed instruments, such as AI contribution form, GenAI integration rubrics, AI contribution assessment form, and GenAI usage instructions, that supported transparent and ethical use.

To truly understand the impact of GenAI integration on students, it had to be systematically monitored and evaluated for which rigorous monitoring and evaluation (M&E) framework was developed, drawing on professional practices in monitoring and evaluation. M&E framework was grounded in theory of change (ToC) that linked input and activities with measurable short-term and medium-term outcomes, capturing both immediate effects after course completion and sustained changes a few months later. Different student cohorts were evaluated for the two courses: project communication and stakeholder management (M562), taught from January to March 2025, which was assessed after completion to measure short-term impact, and management consulting (M563), taught in the summer semester 2024 (April to July 2024), which was assessed several months after completion to measure medium-term impact. The evaluation was conducted across four key dimensions: knowledge, confidence, application, and awareness. The results not only highlighted immediate gains but also showed sustained changes in how students approached GenAI ethically in academic, professional, and personal contexts.

This book therefore brings together practical integration strategies and systematic evaluation insights. It documents in detail how GenAI integration interventions were designed, implemented, and evaluated and makes available practical artifacts such as AI course policies, AI contribution form, GenAI integration rubrics, AI contribution assessment form, GenAI usage instructions, and research instruments, for instance, survey questionnaire, pretesting survey questionnaire, evaluation protocols, and experimentation outcome template, so that educators and institutions can adapt them directly in their own contexts and share GenAI integration experiences with peers.

Our hope is that this book will serve two purposes: first, to inspire educators to see GenAI not as a threat but as a partner in learning through systematic experimentations with structured safeguards, and second, to equip institutions with evidence-based methods for integrating and evaluating GenAI responsibly. By approaching GenAI as a tool for guided exploration, critical reflection, and transparent reporting, higher education can prepare students to thrive in a digital future without compromising academic integrity. The practical and evaluated GenAI integration approaches will offer a clear, ethical, and effective pathway for integrating GenAI into higher education while safeguarding academic integrity.

Potsdam, Germany Varun Gupta
Noida, Uttar Pradesh, India Chetna Gupta

Funding Acknowledgements

The research received financial support from the Faculty Research and Innovation Grant from Gisma University of Applied Sciences, Potsdam, Germany (project ID: RGVG_02, project acronym: DIGEnt).

This publication solely reflects the views of the authors, and the funding agency cannot be held responsible for any use that may be made of the information contained therein. The funder had no involvement in the execution of the study protocol and reporting, including data collection, analysis, decision to publish, or preparation of the manuscript communicating research findings.

Ethical Board Approval

The conducted study was approved by the Institutional Review Board of Gisma University of Applied Sciences, Potsdam, Germany, under the protocol number 010424.

Competing Interests The authors have no competing interests to declare that are relevant to the content of this manuscript.

Contents

About the Authors

Varun Gupta is a Professor of Digital Innovation and Director of the Multidisciplinary Research Centre for Innovations in SMEs (MrciS) at Gisma University of Applied Sciences, Germany, and a Researcher at Universidad de Alcalá, Spain. He holds three doctorates in organizational engineering, economics and business management, and computer science and engineering and has also earned an MBA, a Máster en Dirección Internacional de Empresas, a Master of Technology (By Research), and a Bachelor of Technology in Computer Science and Engineering. His past affiliations include University College London, University of Toronto, Leicester University, Lund University, Sapienza Università di Roma, Free University of Bozen-Bolzano, University of South-Eastern Norway, Poznań University of Technology, and Uniwersytet Szczeciński. His research focuses on software engineering, digital innovation, technology adoption in SMEs, entrepreneurship, and international business management.

Chetna Gupta is a Professor of Computer Science and Engineering with over 20 years of academic and research experience. Her expertise spans software engineering, distributed systems, search-based software engineering, and machine learning. She has supervised multiple Ph.D. and M.Tech scholars and has published extensively in reputed international journals and conferences. Dr. Gupta also serves as an Associate Editor for leading journals and has edited books with reputed publishers. Her current research interests include intelligent bug triaging, human-centric software processes, and innovation management in software enterprises.

Introduction 1

This chapter introduces the integration of generative AI (GenAI) in higher education, setting the foundation for the rest of the book. It first presents the overview of GenAI in academia focusing on opportunities and ethical challenges, highlighting the need for balancing innovation with academic integrity. The chapter then provides a comparative analysis of how three universities—University College London (UCL), Harvard University, and Gisma University of Applied Sciences—are structuring GenAI adoption, emphasizing flexible policies, assessment redesign, and transparent communication with students. Building on the practices of many universities, the chapter proposes a framework that categorizes GenAI integration across courses and assignments into four quadrants: experimental integration, guarded exploration, opportunistic usage, and traditional approach. The GenAI practices of the Multidisciplinary Research Centre for Innovations in SMEs (MrciS) at Gisma University of Applied Sciences fall under experimental integration. This chapter offers a structured lens for understanding the successfully applied approaches to GenAI adoption and sets the foundation for the GenAI integration approaches as applied across different courses, as well as monitoring, evaluation, and impact analysis explored in subsequent chapters.

1.1 GenAI Integration in Academia

The use of generative AI (GenAI) in academic assessments brings both advantages and ethical concerns. While it supports learning (Wang et al., 2023) and strengthens critical thinking (Kizilcec et al., 2024), issues of academic integrity arise when students submit AI-generated work as their own (Cotton et al., 2024; Dwivedi et al., 2023; Lee et al., 2024). To balance innovation with integrity, institutions need to rethink assessment

V. Gupta, C. Gupta, *Technological Innovations in Education*, Synthesis Lectures on Technology Management & Entrepreneurship,
https://doi.org/10.1007/978-3-032-11225-5_1

strategies, enabling students to explore GenAI while reinforcing problem-solving skills and ethical responsibility (Gupta & Gupta, 2024). Traditional assessments focused on knowledge recall are becoming less effective; instead, real-world, application-based tasks should be emphasized (Gruenhagen et al., 2024). Clear guidelines on how to use and acknowledge GenAI in academic work are necessary, especially since research shows that many students do not view AI-assisted work as an integrity issue (Gruenhagen et al., 2024).

To help students gain skills with emerging technology and develop skills to experiment with them, it is important for the institutions to help them experiment with these technologies to explore their capabilities, understand their limitations, and adopt these technologies maintaining academic integrity. In other words, rather than resisting their use in academia, these technologies are required to be integrated in a way that helps students achieve their learning outcomes and maintain academic integrity. This can be achieved by providing students with a controlled experimentation environment during their courses, enabling them to explore the technology under structured safeguards, i.e., protective measures such as rules and controls, so that GenAI can be tested and applied responsibly, without causing unintended harm due to its limitations. The way GenAI is experimented with in courses may vary according to AI course policies, which strongly depend on the nature of the course, intended learning outcomes, and specific regulations in place. For instance, University College London (UCL) has adopted structured approaches to GenAI use, offering a flexible three-tiered framework—ranging from prohibition (Category 0) to assistive (Category 1) and integral use (Category 2) (Gupta & Nyamapfene, 2025a). Harvard University follows a similar decentralized model, allowing faculties to determine AI's role in assessments (Gupta & Nyamapfene, 2025b).

The Multidisciplinary Research Centre for Innovations in SMEs (MrciS) at Gisma University of Applied Sciences, Germany, has implemented an AI course policy for specific courses that defines GenAI as a "knowledge collaborator" in selected courses (Gupta & Nyamapfene, 2025b). To ensure accountability, students must submit an AI Contribution Form and an AI Contribution Assessment Form (or GenAI integration rubrics) with their graded final assessments. Course handbooks and asynchronous and synchronous course tasks are intentionally designed to integrate generative AI (GenAI)-related activities along with traditional methods for generating insights for creative solution formulations. Classroom-based sessions emphasize practical learning through group work, case-based teaching, and student presentations—all incorporating elements of GenAI use and critical reflections about the technology usage. These include guided reflection sessions on the proposed problem solutions, scientific application of research methods using data from diverse sources, and critical evaluations of the technology's outputs. Such activities foster ethical awareness, promote understanding of GenAI's limitations, address intellectual property rights (IPR), and explore data protection concerns, besides other ethical issues with the technology use. This approach encourages students to engage in complex, real-world tasks that strengthen their digital and critical literacy skills while deepening their understanding of GenAI's capabilities, appropriate applications, and potential for misuse.

Assessment briefs, which provide detailed information about the final examination, are discussed with students to clarify expectations and identify areas where GenAI can be appropriately integrated, for instance using GenAI integration rubrics, AI contribution form, and AI contribution assessment form. Students are guided on how GenAI may be used—primarily as a source of direction or inspiration in group work—through detailed usage guidelines. They are also instructed on how such usage should be transparently reported using tools such as GenAI integration rubrics, AI contribution forms, and AI contribution assessment forms. The final examination tasks are intentionally designed to prevent direct resolution through GenAI tools alone. Instead, students must demonstrate their ability to apply classroom learning to solve complex, real-world problems. The evaluation criteria focus on assessing practical aspects such as feasibility, innovation, and implementation of the proposed solutions, as well as the students' ability to justify their decision-making processes—an aspect that is difficult to achieve through technology use alone. Assessment briefs are reviewed and updated regularly to reflect both insights gained from previous cohorts and ongoing advancements in GenAI technologies.

The integration of GenAI into the course required significant changes across multiple dimensions, including course design and delivery, including design of group work, selection of topics for in-depth discussion, redesign of assessment and evaluation rubrics, and development of reporting tools. The overarching goal was to ensure that students not only achieved the established learning outcomes but also actively experimented with GenAI to gain a practical understanding of its capabilities and limitations for identifying the possible application areas and the limits so that technology contributions remain non-substantial.

Drawing on approaches and techniques from the monitoring and evaluation (M&E) professional field, such as the theory of change (ToC), this GenAI integration was structured to link activities with measurable outcomes and impacts. The results from continuous monitoring and formative evaluations of short- and medium-term impacts were conducted, and the results could be instrumental for evolving the GenAI integration approaches. This GenAI integration, as implemented in several courses led by the Director of the Multidisciplinary Research Centre for Innovations in SMEs (MrciS)—who is also one of the authors of this book, the course professor, and module leader, together with the coauthor—was approached as an *intervention*. An intervention is a purposeful action designed to bring about change in a specific target population, in this case, the students. The intervention involved carefully planned inputs, for instance training material; activities, for instance ethics-based discussions on responsible AI use; outputs, for instance reduced number of AI-related academic misconduct cases (AMCs); and short-term, medium-term, and long-term outcomes aligned with monitoring and evaluation (M&E) framework. Regular monitoring and evaluation (M&E) of GenAI's role in education can help track its impact on learning outcomes and ethical behavior (Fernández-Díaz et al., 2017).

The GenAI integration intervention is continuously refined based on insights gained from ongoing course evaluations. These evaluations occur immediately after course completion (immediate evaluation) and several months later (follow-up evaluation). A rigorous monitoring and evaluation (M&E) process assesses both short-term and medium-term outcomes, ensuring continuous improvement and effective implementation of GenAI in education. Throughout the academic course, the intervention was monitored periodically to ensure proper implementation, while evaluation activities aligned with the M&E framework were conducted to measure the impact of GenAI integration across four aspects—knowledge, confidence, application, and awareness.

The GenAI intervention is briefly referenced by Gupta and Nyamapfene (2025b) and is elaborated in detail in this book. It is based on the evaluation of two courses: project communication and stakeholder management (M562), taught from January to March 2025, which was assessed immediately after completion to measure short-term impact, and management consulting (M563), taught in the Summer Semester 2024 (April to July 2024), which was assessed several months after completion to measure medium-term impact. At the time of delivery, these courses were subject to continuous monitoring. Although this book presents the results of short- and medium-term evaluations, the monitored values were very promising. For example, there were zero cases of academic misconduct, whether related to plagiarism or unethical use of GenAI in final assignments (when GenAI was not used as a knowledge collaborator with substantial contributions). Student engagement was high, with greater participation and discussion among group members in classroom activities, on the web-based learning system (Canvas), and in final assignments. Additionally, there was 100% submission of AI contribution forms, even in cases where students used GenAI only to seek guidance rather than to generate assignment outputs. Canvas analytics showed 10,980 page views for the management consulting course (M563) and 11,993 page views for the project communication and stakeholder management course (M562) during the official teaching months of each course, excluding any access beyond the course end date, with 100% of assignments submitted on time for both courses.

This book highlights the GenAI integration in several master's courses (the intervention), provides an in-depth explanation of the M&E framework, and describes the processes used to conduct both immediate and follow-up evaluations alongside the resulting impact of the intervention. The findings presented here aim to assist peer institutions in replicating or adapting these practices, all of which have been validated through rigorous evaluation processes. Additionally, peers will benefit from the detailed M&E framework and evaluation methodologies, enabling them to conduct their own interventions and evaluations internally—rather than relying on third parties—and to leverage evaluation insights to continuously improve their own GenAI interventions. Furthermore, research instruments, such as survey questionnaires, are made available for immediate use or adaptation by other institutions.

To the best of the author's knowledge, there is a notable gap in the literature regarding institutional practices that are systematically evaluated through rigorous M&E processes, particularly in relation to GenAI programs or interventions in academia. This study therefore offers unique and valuable contributions to the academic community.

1.2 GenAI Practices Across Different Universities

Gupta and Nyamapfene (2025a, 2025b) offer an in-depth exploration of how three universities—University College London (UCL), Harvard University, and Gisma University of Applied Sciences—are adopting generative AI (GenAI) in academia. Their comparative analysis highlights not only GenAI practices across academia but also the ways for accelerating its integration and the path ahead for this integration for balancing innovation and responsibility, all grounded in a shared commitment to academic integrity and responsible AI use. The findings are summarized in subsequent paragraphs.

1.2.1 University College London (UCL)

UCL has developed one of the most detailed and structured frameworks for GenAI integration, anchored in a three-tiered classification system that guides faculty on how to incorporate (or restrict) GenAI within assessments. These three categories are not prescriptive but offer flexible, pedagogically informed options for module leaders to choose based on their learning outcomes:

- Category 1 prohibits GenAI usage altogether—except for minor acceptable uses such as grammar checks—ensuring that students meet all learning outcomes independently without AI assistance.
- Category 2 allows GenAI as an optional, assistive tool but does not include GenAI-related skills in learning outcomes or assessment criteria. In such cases, assessments are expected to be inherently GenAI-proof, for example, involving complex, open-ended problems requiring justification of choices by students and inability of technology to generate solutions or evaluate outcomes that GenAI tools cannot directly solve.
- Category 3 integrates GenAI use as a formal part of the assignment, where students are explicitly assessed on their ability to critically and creatively engage with the technology. This includes comparing GenAI outputs to the solutions driven by rigorous research, evaluating their quality, identifying limitations, and reflecting on the reliability and ethical implications of AI-generated content.

In all three categories, UCL places strong emphasis on clearly communicating expectations to students through assessment briefs, aligning decisions with the university's broader policies on feedback, assessment, and academic integrity. Importantly, if GenAI is permitted in an assessment—whether as a tool or a subject of critique—faculty must redesign the assignment to ensure that students' learning outcomes are genuinely evaluated and that academic integrity is maintained. For example, in Category 2, where GenAI use is allowed but not assessed as part of evaluation of the assignment, it becomes essential to embed problems into the assignment that cannot be fully solved using GenAI alone. These problems might require critical analysis, synthesis of interdisciplinary concepts, or justification of optimal decisions, thereby ensuring that students demonstrate independent reasoning beyond what GenAI can offer, limiting the role of technology for seeking directions or automating routine tasks, for instance.

Conversely, in Category 3, where GenAI use is central to the learning task, the assessment design itself must evolve to explicitly evaluate students' ability to scrutinize GenAI outputs. This includes assessing how well students understand the capabilities and limitations of GenAI tools, and whether they can use the technology thoughtfully, ethically, and reflectively. In both scenarios, assessment design becomes the important aspect for a meaningful, responsible integration of GenAI.

UCL does not use GenAI detection tools like Turnitin's AI scoring function, citing concerns about reliability, intellectual property risks, and data privacy. Instead, the institution relies on faculty judgment and, where necessary, investigative vivas to address potential misuse. Supporting this system is UCL's Generative AI Hub, a central platform providing guidance, resources, and toolkits developed by cross-functional working groups, including academic staff, librarians, and digital learning specialists. These resources help educators navigate assessment design, student support, and ethical usage of AI, ensuring that GenAI adoption is accompanied by institutional scaffolding. The platform also provides students with information material about GenAI, for instance how to use the technology, how to acknowledge its use, and ethical issues involved.

1.2.2 Harvard University

Harvard's approach, while sharing UCL's caution, is shaped by its institutional structure and emphasis on secure experimentation. Faculty at Harvard are also given discretion over GenAI use in their courses, guided by overarching principles of data privacy, academic integrity, and compliance with copyright law. Each faculty or school within Harvard—such as the Faculty of Arts and Sciences (FAS)—can select one of the three cases: prohibit GenAI entirely, allow its unlimited use in course assignments, or permit it across few assignments across the course. The permission to use GenAI in the assignment and the extent of the use needs to be communicated to the students through a policy statement in the course (https://oue.fas.harvard.edu/ai-guidance). Importantly, Harvard has invested in dedicated, secure infrastructure to facilitate GenAI use without compromising sensitive information. The AI Sandbox, for instance, allows users to experiment with leading GenAI models from OpenAI, Anthropic, Google, and Meta in a controlled environment (https://huit.harvard.edu/ai-sandbox), with institutional safeguards ensuring that data is neither shared nor used to train third-party models. This sandbox model enables innovation while mitigating risks around data exposure.

Harvard also avoids the use of GenAI detection software due to concerns over accuracy and fairness, echoing UCL's approach. Instead, it also emphasizes assessment redesign to ensure that AI-generated submissions are not undetectably replacing student work. In addition, the "Faculty Voices" initiative showcases reflective narratives from educators experimenting with GenAI, creating a culture of shared learning and peer-led insight.

1.2.3 Gisma University of Applied Sciences

At Gisma, individual module leaders are empowered to determine whether GenAI may be used in assessments. When permitted, its use must be clearly defined in the assessment brief, and students are expected to transparently acknowledge their use of GenAI. For instance, in some of the courses delivered by MrciS director, GenAI was allowed to be used as a "knowledge collaborator" without substantial contributions to the students' work. The course was designed so that various tasks—such as discussions and group work—required the use of GenAI and the research process to synthesize information across different data sources. This approach introduced students to the functionality and limitations of GenAI, as well as to intellectual property and other ethical considerations. It helped students develop critical literacy skills and the ability to formulate evidence-based solutions through rigorous research supported by validated technological outputs. Students engaged with GenAI tools, compared them with manual research findings, and presented their solutions in peer discussions, followed by guided reflection and analysis. Assessment redesign at Gisma reflects an applied, practice-oriented research work. Assessments are intentionally structured so that generative AI tools cannot independently produce complete or meaningful responses. Instead, students are required to synthesize insights from multiple sources, engage in critical analysis, and formulate contextually appropriate solutions. In this model, the role of generative AI could be restricted to tasks like providing support for understanding complex terminology, generating assignment outlines, enhancing academic language, or offering directional guidance—rather than as a substitute for analytical thinking or subject mastery. Students are obligated to report the level of the technology's contribution, describe their technology usage, and acknowledge its use.

The impact of generative AI (GenAI) integration on students is continuously monitored and evaluated through a rigorous process aligned with an established monitoring and evaluation (M&E) framework. Both short-term and long-term outcomes are assessed using structured M&E instruments designed to capture changes in student knowledge, confidence, application, and awareness aspects. The integration of generative AI (GenAI) into courses has been guided by a clearly articulated theory of change (ToC), which outlines the key activities, outputs, and intended outcomes. The ToC maps the pathway through which the integration of GenAI is expected to generate measurable impact on students, ensuring a coherent link between intervention design, implementation, and evaluation. The integration of generative AI (GenAI) in some of these courses, along with the monitoring, evaluation, and learning (M&E) of its short- and long-term impacts, is detailed in Chaps. 2, 3, 4, and 5 of this book.

1.2.4 Cross-Institutional Synthesis of GenAI Integration Strategies: UCL, Harvard, and Gisma

In analyzing the GenAI integration strategies of UCL, Harvard, and Gisma, several overarching themes become evident highlighting not only institutional commitment to responsible adoption, but also a shared understanding of the pedagogical and ethical complexities involved. Despite contextual differences, all three universities converge on a core principle: faculty should have the flexibility to decide whether and how GenAI is incorporated into final assignments, with the expectation that student learning outcomes must be meaningfully and independently demonstrated. From this foundation, Gupta and Nyamapfene (2025a, 2025b) identify several key themes that underpin GenAI-related practices across institutions:

- First, there is a strong emphasis on redesigning assessment briefs to maintain academic integrity. Where GenAI is permitted, assignments are restructured to either challenge the tool's capabilities—such as by embedding critical or contextual problems that GenAI cannot adequately address—or assess the student's ability to critically evaluate GenAI outputs, especially when GenAI usage is itself a learning outcome. This design logic ensures that students remain central to the learning process and are not simply outsourcing intellectual effort to technology. In other words, GenAI-generated content is not assessed in student assignments and what is evaluated is their ability to show their creativity to formulate an effective solution to the problem and integrate GenAI technology ethically like other technologies without fully relying on their outcomes. By this, the student's ability to meet the intended learning outcomes, apply critical problem-solving skills, and use the technology ethically is central to the evaluation process. Merely permitting the use of GenAI in assessments is not sufficient. Its inclusion must be supported by thoughtful assignment design. If GenAI is intended to play an active role, the task must require students to critically engage with the tool, assess the quality and relevance of its outputs, and demonstrate a level of understanding that goes beyond simple prompting. Alternatively, if GenAI use is optional or assistive, the assignment should incorporate components—such as the justification of complex decisions or the synthesis of diverse ideas—that require students to exhibit originality, analytical reasoning, and intellectual depth beyond the tool's current capabilities.
- Second, universities focus on clear and transparent communication with students. Faculty are expected to state explicitly—within assessment briefs and syllabi—whether GenAI use is allowed, to what extent, and under what conditions. Boundaries are clearly drawn to prevent misunderstanding and to ensure that students are aware of the expectations and responsibilities related to GenAI use.
- Third, institutions have developed dedicated online spaces or internal platforms to guide both staff and students in engaging with GenAI tools responsibly. These platforms, such as UCL's Generative AI Hub, offer practical support including policy

overviews, ethical guidelines, use-case examples, training materials, and instructions for proper acknowledgement of GenAI contributions.

- Fourth, where GenAI experimentation is encouraged, secure environments are provided to minimize risk. Harvard's AI Sandbox and UCL's use of Microsoft Copilot (under institutional licensing) exemplify this approach—allowing staff and students to explore GenAI in a controlled setting without jeopardizing data privacy or intellectual property.
- Fifth, there is a growing emphasis on peer learning and community engagement. Faculty are encouraged to share their experiences, reflect on course redesigns, and contribute to evolving best practices. Initiatives such as Harvard's "Faculty Voices" or the reflective culture fostered at Gisma through weekly staff meetings demonstrate how institutions are leveraging practitioner insights to inform institutional learning.
- Sixth, while traditional plagiarism detection software like Turnitin remains in use, there is minimal reliance on GenAI-specific detection tools. Concerns around the accuracy, fairness, and data practices of such tools have led these universities to instead rely on faculty judgment, assessment design, and investigative processes (such as vivas) to uphold academic integrity.
- Seventh, all three institutions highlight the importance of interdisciplinary working groups—comprising educators, librarians, digital learning specialists, legal experts, and IT staff—to co-create policies and provide holistic support. These groups play a critical role in adapting institutional practices to keep pace with the rapid evolution of GenAI technologies.
- Finally, there is a strong culture of continuous improvement and feedback. Institutions treat GenAI integration as an ongoing learning process, gathering evidence from faculty experiences with GenAI integration in their courses, assessment outcomes, and short-term and long-term impact evaluations *(for instance, those employed at MrciS)* to refine their strategies. Policies (or better call them guidelines) are not static—they are iteratively improved based on what works in practice, ensuring responsiveness to both technological advances and pedagogical insights.

1.3 Integration of GenAI in University Courses: A Framework

Based on authors' observations and experiences across multiple universities (or at the course level for universities with supporting GenAI policies), including authors' own teaching experiences, GenAI integration appears to follow four distinct patterns based on two factors, *course-wide integration (whether GenAI is embedded throughout the course) and assignment-level usage (whether its use is permitted in student-graded assessments).* Where integration is allowed or not, the tutor determines the extent of incorporation, which may range from low to high levels, provided that the institutional guidelines allow it. McDonald et al. (2025) analyzed publicly available GenAI policy documents from 151 universities and found that 27% discourage GenAI use, with 9% offering minimal guidance. Most (63%) encourage its use, of which 41% provide detailed guidance while 22% do not. The remaining 10% were considered neutral. Discourage, as per the authors, means universities signal that GenAI is unwelcome in classrooms, either through curricula designed against its use or by emphasizing its legal and ethical risks. Encourage, as per the authors, means universities do not discourage GenAI and provide substantial guidance on its use in the classrooms, or related activities, such as lesson planning and clearly communicating ethical GenAI policies to students.

In the context of university-level education, the integration of generative AI (GenAI) into teaching and assessment practices can be understood through two key factors:

- **Course-wide integration:** It signifies whether GenAI is integrated across the entire course. In other words, it reflects if GenAI is embedded into the course design, including learning activities, discussions, and student training.
- **Assignment-level usage:** It signifies whether GenAI usage is permitted in graded assignments. In other words, it represents whether students are allowed—or even encouraged—to use GenAI tools in graded assessments, including both intermediate tasks and final submissions.

These two dimensions create a framework with four distinct quadrants, each representing a different pedagogical approach to engaging with GenAI (Fig. 1.1). Each has different intentions, practical implications, and expectations for students and faculty. These quadrants are not hierarchical but strategic—each has its own rationale depending on the learning goals, student cohort, course context, and institutional guidelines. Figure 1.1 graphically shows the four different GenAI integration approaches of various universities.

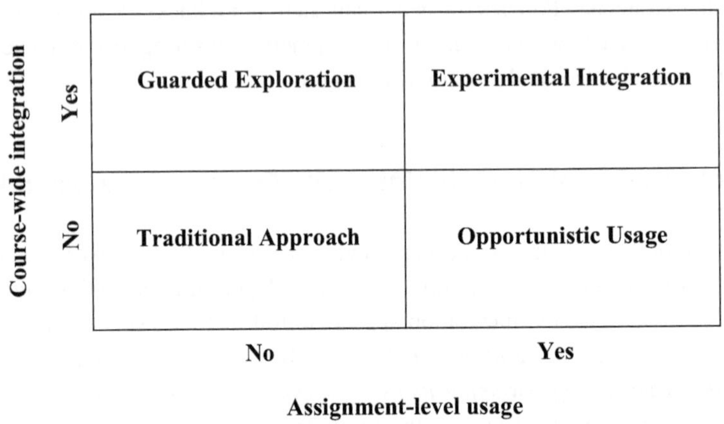

Fig. 1.1 Framework for GenAI integration in courses based on course integration and assessment usage

1.3.1 The Four Quadrants of GenAI Integration

Let us now consider the four quadrants created by the two dimensions—course-level integration and assignment-level usage. Each quadrant has different pedagogical intents, practical implications, and expectations for students and faculty.

(a) **Experimental Integration (course integration and permitted in assignments)**
In experimental integration, the courses embrace GenAI as a pedagogical tool throughout the semester and in final assessments. GenAI is introduced early, students are trained in its use during the course, and course activities are designed to encourage experimentation and reflection. Final assignments allow GenAI as a knowledge collaborator, with structured expectations around ethical use and transparency. Assignments can be designed either to make it difficult or impossible for students to rely on generative AI as a substitute for their own work or to intentionally incorporate responsible AI use as part of the evaluation. For example, assessments could focus on students' ability to critically reflect on the functionality and limitations of the technology when applied to their practical assignments.
In the courses mentioned in this book, GenAI was fully integrated throughout the course. Students receive structured training not only in using the technology but also in understanding its functionalities, limitations, and ethical concerns. This is done through changes in course design and delivery, course evaluation adaptations, and assessment redesigns (*for more details, please refer Chap. 2*). This approach prepares students to engage with generative AI critically and strategically under the tutor's guidance, while peer learning fosters further experimentation with the technology, both collaboratively and individually. Final assignments in such courses permit the use of GenAI as a knowledge collaborator, with clear guidelines on when and how to use it responsibly. At the institutional level, the prevailing guidelines give faculty autonomy to determine whether GenAI can be used in their courses and, if so, to what extent.

(b) **Guarded Exploration (course integration only, but not permitted in assignments)**
Guarded exploration involves the exploration of GenAI throughout the course—students learn about the tool, engage in practice activities, and reflect on its capabilities and risks, but its use is restricted or even prohibited in final assignments. This model emphasizes literacy and critical awareness over application in summative tasks. The aim is to foster critical understanding and awareness rather than practical application in graded assessments. Students could be motivated to use the technology for nongraded assignments. This approach supports AI literacy while maintaining traditional assessment formats to preserve originality or discipline-specific expectations. For instance, some universities may still evaluate students through traditional, in-class exams completed on paper without the use of technology. Moreover, the nature of certain courses—such as foundational or language courses—may require restricting the use of technology in graded assignments.

One example could be of a foreign language learning course. Here, GenAI is integrated into classroom activities to help students practice writing, analyze grammar and vocabulary, and reflect on which AI-generated outputs are correct or flawed. However, in the final graded exam or essay, GenAI use is prohibited to ensure that students can demonstrate independent language proficiency.

(c) **Opportunistic Usage (No course integration but permitted in Assignment usage)**

Opportunistic usage involves GenAI not explicitly taught or discussed during the course itself, but students are permitted—or even expected—to use it in assignments. This model assumes that students have prior exposure to GenAI from earlier coursework, self-study, or institutional resources. Faculty trusts students to make appropriate use of AI tools in support of their learning and outputs, without embedding GenAI instruction into the course structure. In this approach, again assessments need to be carefully designed to ensure that AI cannot replace student effort, critical thinking, or creativity to maintain academic integrity.

One example could be that of the master's thesis module. In such module, GenAI could not be integrated into the teaching of the module, for instance research workshops, but its use is permitted in thesis development tasks such as refining drafts, structuring outlines, or brainstorming research directions. Since students already had structured exposure to GenAI in earlier courses, they are trusted to apply it responsibly with safeguards like an AI policy, AI contribution form, AI contribution assessment form, and GenAI integration rubric outlined in the thesis handbook.

(d) **Traditional Approach (No course integration and not permitted in assignments)**

Traditional approach is the most conservative approach; neither the course content nor assignments involve GenAI. Either the topic does not lend itself to GenAI integration or the faculty has chosen to prioritize conventional learning and assessment methods. Although still prevalent in some universities, this approach was particularly common when ChatGPT was first launched, as the education sector initially took time to respond to the innovation.

One good example here, apart from the courses taught in universities that are still figuring out their GenAI-related approaches, could be that of the clinical psychology practicum. This is a practical, supervised, real-world training experience where students apply classroom learning by seeing clients for therapy, conducting psychological assessments, and participating in clinical activities under the direct guidance of a licensed psychologist. Because the practicum involves highly sensitive personal data, confidential case notes, and first-hand clinical observations, GenAI will never be integrated into the course design. Nor will it ever be permitted in final assignments, since uploading or processing such information in external AI systems would breach patient confidentiality, violate ethical codes, and contravene data protection regulations such as GDPR. The emphasis is instead on developing professional judgment, empathy, and applied clinical competence that cannot be delegated to technology.

To foster the integration of generative AI (GenAI) in the course, practical experiences with the courses discussed in the book and detailed in Chap. 2 suggest several recommended changes to course design and delivery. These include the following:

- **Enhancing collaborative group work**: Weekly, challenging group activities encourage students to engage actively with technology, develop digital literacy skills, and integrate validated technological outputs into optimal solutions. This entails providing detailed technology usage guidelines that not only outline solutions but also document technology outputs—both validated and erroneous—along with reflections on the reasons behind them, across various capabilities and limitations, and strategies to address these limitations.
- **Leveraging detailed analysis materials:** Both open-access files (to explore technology analysis capabilities) and copyrighted materials (to examine copyright issues) are used in group work, allowing students to critically analyze different content types and apply technology appropriately.
- **Expanding research topics in the module handbook:** Including underexplored, applied research topics fosters creativity, encourages the use of multiple data sources, and promotes experimentation with diverse technologies.

To foster the integration of generative AI (GenAI) in graded assignments, practical experiences with the courses described in the book and detailed in Chap. 2 recommend several changes in assessment design and evaluation. These include the following:

- **Explicit AI policy:** Courses that permit GenAI use clearly state the policy in the course documentation, outlining whether usage is allowed, the expectations and concerns regarding its use, and the required documentation that students must submit if they incorporate technology in their final assignments.
- **Applied research-focused assignments**: Assignments are designed around open-ended, applied research problems to encourage creativity and critical thinking.
- **GenAI integration rubric**: A rubric is provided for final assignments in each course to guide students on the role of technology in research work, highlighting activities where its use is appropriate, situations where its use would be unethical, and contribution levels for self-reporting.
- **Ethical technology use and transparency**: Students are required to submit the following documents to document GenAI's role as a knowledge collaborator:
  - AI contribution form for reporting the technology usage in the graded assignment research work
  - AI contribution assessment form (or GenAI integration rubric form) for reporting the technology role in research process and self-evaluation of contribution levels ensuring non-substantial contribution of the technology
- **Alignment with research focus**: Assignment rubrics are closely aligned with the applied research emphasis of the final assignments, ensuring that GenAI supports

learning and idea development without directly influencing the assessment of practical, skill-based evaluation criteria.

By mapping GenAI integration across these four GenAI integration approaches (the four quadrants of the framework), universities and faculty can design more intentional and pedagogically aligned experiences. Students benefit most when expectations are clear, rationale is transparent, and opportunities for dialogue and reflection are embedded throughout the course. The courses mentioned in this book follow experimental integration approach for GenAI integration involving students experimenting with GenAI during the learning process and have robust guidance when it comes to using AI in their final practical and challenging assessments. The integration of generative AI in the courses involves multiple safeguards, including controlled experimentation guided by rigorous GenAI use instructions; clear institutional AI policy; mandatory reporting and acknowledgement of AI use through AI contribution form, AI contribution assessment form, and GenAI integration rubrics; and measures to ensure that its contribution remains non-substantial *(for more details, refer to Chap. 2 of this book).*

When GenAI usage and integration are implemented, they will follow a scale from low to high. If GenAI is permitted in final assignments, integrated into the course, or both, students should be provided with a clear AI policy outlining the allowed level of contributions, what constitutes unethical use, and the procedures for reporting and acknowledging their use of the technology *(recommendations for assessment redesign and evaluation adaptation).* Furthermore, previously recommended changes in course design and delivery should be implemented to accommodate the intended level of engagement with the technology.

This book presents approaches for courses with substantial GenAI integration and the use of GenAI in graded assignments as a knowledge collaborator rather than as a substitute for student work *(non-substantial GenAI contribution).* Students are free to use GenAI extensively; however, due to the practical and open-ended nature of the assignments, the technology cannot replace their own contributions, and the graded assignments ensure that the work remains AI-proof. The courses are designed to allow experimentation with GenAI to explore its functionality and limitations, develop strategies to address these limitations, and provide opportunities for students to report both their use of the technology and its non-substantial role. These practical lessons are valuable for any course planning to permit GenAI usage in assignments and/or to integrate it into the course design, including learning activities, discussions, and student training.

1.4 Research Objectives

This book is driven by two primary, interconnected objectives that respond directly to the pressing need for actionable, evidence-based guidance on integrating generative AI in higher education.

First, the book aims to disseminate a detailed, practice-based account of how GenAI can be purposefully integrated into academic courses. This includes an in-depth examination of how the final assessments were redesigned to make it impossible for students to rely solely on generative AI for solutions. Instead, the redesigned tasks required students to engage in deeper research and demonstrate advanced literary and critical thinking skills. Furthermore, the AI policy for the course was drafted to provide clear guidance to students on whether generative AI is permitted, and if so, how they can use it ethically, how they should report their actual use of the technology, and what limitations they need to be mindful of. This also includes supporting students through clear instructions and structured artifacts—such as AI contribution forms—which guided them in disclosing when and how they used generative AI in their assignments. The teaching across multiple sessions was carefully designed to help students use technology as a collaborator rather than a substitute, emphasizing that generative AI should be treated as just one source of information. Students were encouraged to critically validate and enrich AI-generated outputs by triangulating them with data from other reliable sources. The improvement of students' literary skills, combined with a better understanding of the capabilities and limitations of the technology, will help them identify appropriate use cases as well as potential misuse cases. By sharing the GenAI integration practices in full detail, the book provides practical guidance that other educators and institutions can adapt to promote ethical, transparent, and pedagogically meaningful GenAI use.

Second, the book aims to equip peers with a robust framework for monitoring and evaluating the impacts of GenAI integration in their own contexts. To this end, the book documents the entire monitoring and evaluation (M&E) framework used to assess the short-term and medium-term outcomes of the intervention across four key aspects: knowledge, confidence, application, and awareness. It includes the detailed theory of change (ToC), survey instruments, pretesting processes, and evaluation process. By making these M&E artifacts openly available, the book supports peer institutions in reusing, adapting, or extending these tools to measure the impacts of their own GenAI interventions, whether in individual courses or across wider programs.

Together, these two objectives address a critical gap in the current literature and practice: although many institutions are experimenting with GenAI, few have systematically documented how it can be integrated and how its impacts can be rigorously evaluated and refined over time. By filling this gap, this book aims to support the wider academic community in building not only AI fluency but also sustainable, ethical, and transparent GenAI practices.

1.5 Contributions of the Book

This book addresses a significant gap in both academic literature and institutional practice regarding the integration of generative artificial intelligence (GenAI) in higher education. It offers a comprehensive, experience-based account of how integrating generative AI

(GenAI) into academic courses is not a matter of simply allowing access to a tool; it requires a thoughtful and systemic change. For GenAI to be used ethically and meaningfully, courses must be redesigned and delivery mechanisms modified to incorporate it purposefully into learning outcomes and activities, for instance by providing technology usage instructions, challenging group tasks to help students explore technology and address their limitations, and tools like GenAI integration rubrics. The redesign process includes guided experimentation, reflection, and collaborative learning. Additionally, assessments must be restructured to ensure that GenAI supports, rather than replaces, student effort—emphasizing real-world application, critical thinking, and originality. The redesign ensures assessing students' ability to strengthen their information literacy, critically engage with GenAI as a knowledge collaborator, and formulate justifiable, research-informed solutions. This means evaluating not only how students use GenAI, but also how effectively they integrate its outputs with evidence from credible sources and apply critical thinking to synthesize and present well-reasoned responses. Finally, evaluation strategies must also evolve, providing students with new instruments that allow them to self-assess the GenAI contribution in their assignments, and self-report the technology usage, for instance using AI contribution form, GenAI integration rubrics, and AI contribution assessment form. This also involves selecting evaluation rubrics that align with the practical aspects of the assignment, which are more challenging for GenAI to fully address. This book addresses each of these dimensions with practical tools and empirically grounded insights. Drawing on a series of implemented and evaluated interventions in postgraduate courses, the book makes distinctive contributions across theoretical, methodological, and practical dimensions.

1.5.1 Theoretical Contributions

At the conceptual level, the book introduces a structured pedagogical and evaluative framing for integrating GenAI into academic coursework, which has been largely absent in the existing literature.

- **Theory of change (ToC) for GenAI in education**: The book provides a custom-developed theory of change model specifically for GenAI integration in academic settings. It outlines the logical connections between intervention strategies, inputs, outputs, and intended outcomes, providing a foundational framework for understanding how and why GenAI use in classrooms leads to meaningful learning and behavior change in short and long time periods *(called Impacts).*
- **Framing GenAI as a "Knowledge Collaborator"**: The book advances the idea that GenAI should not be viewed solely as a threat to academic integrity or as an automation tool, but as a knowledge collaborator, provided that changes are made in courses, assessments, and evaluations accordingly. The use of technology as a knowledge collaborator means employing it to obtain guidance, directions, or support during research

or assignments without allowing it to make a substantial contribution to the final work. Ethically, the AI should not be considered a co-author or being eligible for it. This clearly signifies that academia should be positive toward the use of the technology, but students will have to realize that technology cannot be the substitute for their own efforts to ensure academic integrity. The students and technology can co-create knowledge ensuring transparency, ethical usage, and reflective learning.

- **Expanding outcome dimensions**: The book proposes and operationalizes four dimensions of GenAI-related learning outcomes: knowledge, confidence, application, and awareness. These aspects are used in real settings, for instance in American Library Association (ALA) Libraries Build Business (LBB) initiative impact evaluation (**American Library Association,** 2022, 2023), and hence reusing them for GenAI impact evaluations provides a practical, validated framework to assess how learners acquire knowledge, build confidence, apply skills, and develop awareness of responsible and effective AI use. The outcomes, indicators, and underlying survey questions for evaluating the impact of GenAI on students differ from those selected for the ALA Libraries Build Business (LBB) initiative.

These theoretical contributions lay the groundwork for designing, implementing, and evaluating GenAI use in academia in ways that are pedagogically sound and ethically informed.

1.5.2 Methodological Contributions

The book makes significant contributions to the evaluation and research methodology of GenAI technology integration, where such work remains scarce in the literature.

- **Rigorous monitoring and evaluation (M&E) framework**: The book details the full M&E framework applied to GenAI integration interventions. It includes a structured design for both immediate (short-term) and follow-up (medium-term) evaluations using the knowledge, confidence, awareness, and application dimensions. This dual-phase evaluation helps capture not only initial learning but also sustained behavioral change and ethical awareness over time.
- **Development and pretesting of research instruments**: The book outlines the creation and pretesting of survey instruments used to assess student outcomes. The process included validating the survey instrument against aspects like clarity and comprehension of questions, appropriateness and acceptability of response options, impartiality and lack of bias, overall quality of the questionnaire, clarity and sufficiency of instructions, logical flow of questions, and time required to complete the survey. The survey instrument was improved based on feedback from diverse student groups that participated in the pretesting. This ensures that the instruments are valid and accurate for the survey-based evaluation to be conducted with the students.

- **GenAI integration practices grounded on real practices:** The GenAI integration practices detailed in Chap. 2 are not theoretical suggestions—they stem directly from real-world implementation in multiple postgraduate courses. These practices were applied in actual classroom settings, involving diverse student cohorts, and were subjected to rigorous monitoring and evaluation. As a result, the insights and strategies presented are grounded in lived teaching experience and supported by practical, observable outcomes aligning with the theory of change (ToC) of M&E framework. This applied foundation ensures that the guidance offered in the book is not only pedagogically sound but also realistically adoptable by peers in similar academic environments.
- **Mixed-methods data collection and analysis during the impact evaluation**: Combining quantitative survey data with qualitative open-ended responses during the evaluation, the book employs a mixed-methods approach that enables deeper interpretation of the impact of GenAI on students in the short and long terms. This approach adds methodological robustness and enhances the explanatory power of the findings.
- **Reusable and validated artifacts**: Reusable and validated artifacts were created through the professional experiences of the course instructor and actual classroom practice. They serve as methodological tools for both researchers and educators seeking to replicate or adapt the GenAI integration in their courses and the evaluation framework. These artifacts include the following:
  - AI policy and assessment briefs
  - AI contribution form
  - AI contribution assessment forms
  - GenAI integration rubrics
  - GenAI usage instructions for classroom group tasks (could be optimized for any tasks)
  - Theory of change model
  - Pretesting protocols and feedback loops
  - Survey instruments for immediate and follow-up evaluation

1.5.3 Practical Contributions

Beyond theory and research, the book offers highly practical insights for educators, course designers, and academic institutions interested in integrating GenAI in a responsible and effective manner.

- **Course-level integration practices**: The book presents in detail how GenAI was integrated into multiple postgraduate courses. This includes modifications to course delivery, group tasks, reflective exercises, and assessments designed to promote ethical experimentation with GenAI that helps students meet their learning outcomes and ensures academic integrity, along with technology usage instructions and usage reporting forms. These practices can be directly adopted or adapted by other institutions.

- **Ethical use and disclosure practices**: Practical mechanisms for ensuring ethical usage—such as providing GenAI usage instructions to help students explore technology considering different limitations and understanding strategies to address those limitations. This also requires students to submit AI contribution form, GenAI integration rubrics, and AI contribution assessment form and reflect on their use of GenAI—whether it is documented and assessed. These practices not only safeguard academic integrity but also cultivate responsible technology use and honest reporting of the technology contributions through acknowledgements among students.
- **Demonstrated impact on learning and behavior**: The book provides evidence of reduced academic misconduct cases, improved student understanding of GenAI's capabilities and limitations, and higher knowledge, confidence, application, and awareness in using GenAI across academic and professional contexts—even several months after course completion. This long-term impact supports the practical value of integrating GenAI thoughtfully.
- **Support for institutional adoption**: Peer institutions can benefit from the ready-to-use materials, tested methodologies, and adaptable GenAI integration practices presented in the book. These elements offer a scalable starting point for broader institutional or program-level integration of GenAI.
- **Capacity building for faculty**: By demonstrating how to redesign courses and the delivery, redesigning GenAI-inclusive assessments, adapting course evaluations, and conducting M&E in-house without third-party M&E experts, the book supports professional development and institutional capacity building.

This book offers a rare and valuable contribution at the intersection of theory, method, and practice. It not only helps educators and institutions navigate the ethical and pedagogical challenges of GenAI, but also equips them with the frameworks, tools, and evidence to do so with clarity and confidence. By addressing both how GenAI can be integrated and how its impact can be meaningfully measured, this work stands as a resource for those seeking to foster AI-literate, ethically aware, and industry-ready graduates.

References

American Library Association. (2022). *Libraries build business playbook.* Retrieved August 01, 2024, from https://www.ala.org/sites/default/files/advocacy/content/Workforce/LBB_Playbook_web_020222.pdf

American Library Association. (2023). *Libraries build business.* Retrieved August 01, 2024, from https://alair.ala.org/server/api/core/bitstreams/2e276162-edb2-4fbc-9d2b-26ba8eb43691/content

Cotton, D. R., Cotton, P. A., & Shipway, J. R. (2024). Chatting and cheating: Ensuring academic integrity in the era of ChatGPT. *Innovations in Education and Teaching International, 61*(2), 228–239. https://doi.org/10.1080/14703297.2023.2190148

Dwivedi, Y. K., Kshetri, N., Hughes, L., Slade, E. L., Jeyaraj, A., Kar, A. K., et al. (2023). Opinion paper: "So what if ChatGPT wrote it?" multidisciplinary perspectives on opportunities, challenges and implications of generative conversational AI for research, practice and policy. *International Journal of Information Management, 71*, 102642. https://doi.org/10.1016/j.ijinfomgt.2023.102642

Fernández-Díaz, M. J., Rodríguez-Mantilla, J. M., & Jover-Olmeda, G. (2017). Evaluation of the impact of intervention programmes on education organisations: Application to a quality management system. *Evaluation and Program Planning, 63*, 116–122. https://doi.org/10.1016/j.evalprogplan.2017.04.005

Gruenhagen, J. H., Sinclair, P. M., Carroll, J. A., Baker, P. R., Wilson, A., & Demant, D. (2024). The rapid rise of generative AI and its implications for academic integrity: Students' perceptions and use of chatbots for assistance with assessments. *Computers and Education: Artificial Intelligence, 7*, 100273. https://doi.org/10.1016/j.caeai.2024.100273

Gupta, V., & Gupta, C. (2024). Navigating the landscape of AI-generated text detection: Issues and solutions for upholding academic integrity. *Computer, 57*(11), 118–123. https://doi.org/10.1109/MC.2024.3445068

Gupta, V., & Nyamapfene, A. (2025a). Generative AI in universities: Practices at UCL and other institutions, and the path forward. *Internet Reference Services Quarterly, 29*(1), 1–21. https://doi.org/10.1080/10875301.2025.2453461

Gupta, V., & Nyamapfene, A. (2025b). Integrating generative AI in academic assessments: Practical recommendations on AI usage and enhancing knowledge sharing. *Internet Reference Services Quarterly, 29*, 1–11. https://doi.org/10.1080/10875301.2025.2456903

Kizilcec, R. F., Huber, E., Papanastasiou, E. C., Cram, A., Makridis, C. A., Smolansky, A., et al. (2024). Perceived impact of generative AI on assessments: Comparing educator and student perspectives in Australia, Cyprus, and the United States. *Computers and Education: Artificial Intelligence, 7*, 100269. https://doi.org/10.1016/j.caeai.2024.100269

Lee, D., Arnold, M., Srivastava, A., Plastow, K., Strelan, P., Ploeckl, F., Lekkas, D., & Palmer, E. (2024). The impact of generative AI on higher education learning and teaching: A study of educators' perspectives. Computers and Education: Artificial Intelligence, 6, 100221. https://doi.org/10.1016/j.caeai.2024.100221

McDonald, N., Johri, A., Ali, A., & Collier, A. H. (2025). Generative artificial intelligence in higher education: Evidence from an analysis of institutional policies and guidelines. *Computers in Human Behavior: Artificial Humans, 3*, 100121. https://doi.org/10.1016/j.chbah.2025.100121s

Wang, S., Sun, Z., & Chen, Y. (2023). Effects of higher education institutes' artificial intelligence capability on students' self-efficacy, creativity and learning performance. *Education and Information Technologies, 28*(5), 4919–4939. https://doi.org/10.1007/s10639-022-11338-4

GenAI Integration in Courses

2

This chapter presents how generative AI (GenAI) was integrated into postgraduate courses, with a focus on strategies that maintain academic integrity while enabling structured experimentation. It contrasts reactive measures, such as disciplinary actions after misconduct, with proactive approaches that include AI policies, usage guidelines, and faculty-supported exploration. The integration is structured across three approaches: course redesign and delivery, assessment brief redesign, and course evaluation adaptations. Practical approaches include redesigning group activities into applied, research-driven tasks; providing detailed GenAI usage instructions; requiring acknowledgement and reporting tools such as AI contribution forms, GenAI integration rubrics, and AI contribution assessment form; and adapting evaluation rubrics to ensure the inability of technology to act as a substitute for the student work. Through guided experimentation in these areas, students were able to explore GenAI's capabilities and limitations, reflect on appropriate and inappropriate use cases, and strengthen their digital literacy, reflective synthesis, and ethical awareness. These integration practices, implemented across multiple courses, are discussed in depth as a structured intervention.

2.1 GenAI Integration in Academia: An Introduction

In courses adopting AI technologies, strategies to maintain academic integrity can be classified as either proactive or reactive (Gupta & Gupta, 2024). Reactive methods, such as disciplinary actions following cases of academic misconduct, address violations only after they occur. Proactive strategies, on the other hand, aim to prevent misconduct by providing AI training, establishing clear usage policies, and fostering ethical engagement with emerging technologies.

V. Gupta, C. Gupta, *Technological Innovations in Education*, Synthesis Lectures on Technology Management & Entrepreneurship,
https://doi.org/10.1007/978-3-032-11225-5_2

When the objective is to align students with industry practices by encouraging experimentation with new technologies, it becomes essential to enhance their learning through faculty-supported exploration. Continuous guidance throughout the course enables students to test the functional capabilities of AI while also recognizing its limitations and developing strategies to overcome them. This process reinforces the understanding that technology cannot substitute for students' own intellectual efforts; rather, it should be employed as a supporting tool to complement traditional research approaches in generating creative solutions to problems.

Through this guided experimentation, students not only identify possible use cases but also reflect on potential misuse cases of AI. Furthermore, by leaving space for self-directed experimentation, they are empowered to discover more advanced applications over time, benefiting from ongoing innovations within AI as well as in related technologies. This requires a focus on proactive strategies that incorporate generative AI (GenAI) into courses from day 1. Such integration enables students to combine their research-driven insights with technology-enabled outcomes, helping them to fully grasp the value of the technology while becoming more aware of academic integrity.

To achieve this, assessment briefs also need to be redesigned, shifting toward applied, research-based assignments that require critical engagement rather than tasks easily completed through GenAI. In addition, an AI course policy is required to be provided to the students at the beginning of the course. The AI course policy should clearly specify whether GenAI use is permitted in the course. If allowed, it must outline the extent of use, the areas of the assignment that can benefit from the technology—potentially documented through a GenAI integration rubric—and the limitations of the technology. The policy should also define how students are expected to acknowledge GenAI contributions, ensure that its role remains as a supportive knowledge collaborator without substantial contribution to maintain academic integrity, and report any significant interactions with the technology that shaped the assignment. By providing this guidance, the policy fosters transparency and ethical use, enabling students to integrate GenAI responsibly into their academic work. Additionally, instructional material should be provided that explains the possible roles GenAI can play in course assignments, for instance *AI contribution form*, along with examples of assignment-related tasks aligned with those roles (e.g., clarifying whether GenAI can support ideation, literature scanning, drafting, or critique, but not substitute core student work, for instance using CRediT roles), for instance, *GenAI integration rubric form*. Assessment redesign is mandatory in the era of GenAI (Gruenhagen et al., 2024; Gupta & Gupta, 2024). This is because students may submit assignments generated by GenAI as their own work, thereby compromising academic integrity (Cotton et al., 2024; Dwivedi et al., 2023; Lee et al., 2024; Kizilcec et al., 2024).

Gupta and Gupta (2024) reported that the absence of precise AI detection solutions, rise of complementary AI humanization technologies, students' adoption of humanizing tactics, and legal concerns following detection collectively pose significant challenges to identifying AI-generated text. To address these issues, several solutions have been proposed. These include designing more applied research and practice-based assignments developed with guidance from faculty and other stakeholders; incorporating CRediT statements to explicitly disclose the role of GenAI and ensure that its contribution remains minimal; and using research protocols to increase transparency in the research process. Together, these approaches provide more robust mechanisms to safeguard academic integrity while still allowing students to benefit from responsible experimentation with GenAI.

Additionally, course evaluation processes should also be adapted. This includes requiring students to disclose and acknowledge the use of GenAI in final assignments, ensuring that its contribution remains non-substantial, for instance using *AI contribution form and GenAI integration rubric form*. Evaluation rubrics should be aligned with the practical and creative nature of the work, focusing on tasks that cannot be directly solved by GenAI alone.

Just ensuring assessment redesign and course evaluation process adaptation will not sufficiently help students explore the potential of the technology. Assessments and evaluations are the last stages of the course. Redesigning assessments and evaluation adaptations alone will be like merely setting expectations, providing some instructions, and then leaving students to self-experiment and learn the ethical use of the technology on their own.

Exploring the technology to find its practical application aligning with the academic integrity is a journey that requires guiding students through multiple stages, where they acquire knowledge, build confidence with the technology, develop awareness, and apply their learning into practice that reflects the realities of the technology. Achieving this requires a pedagogical approach that embeds GenAI in learning activities, compelling students to combine it with other technologies and traditional methods to generate insights and develop solutions. This creates a controlled learning environment, and within this setting, they are encouraged to formulate strategies to overcome these limitations, thereby developing a deeper and more critical understanding of GenAI as a supportive tool rather than a replacement for their own intellectual contributions. Course redesign involves introducing more complex and challenging group work, supported with detailed guidance on how GenAI may (and may not) be used, besides other activities exposing students to technology realities.

The courses discussed in this book ethically integrate GenAI across three areas: course redesign, assessment redesign, and evaluation adaptations, using various mechanisms outlined in the following sections. Figure 2.1 graphically represents the areas along with different approaches used.

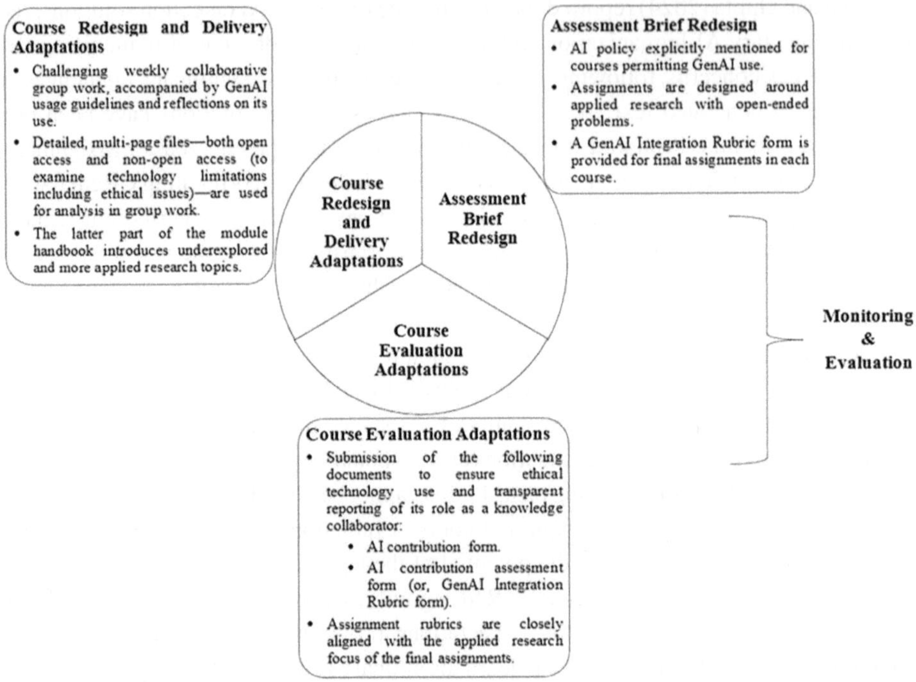

Fig. 2.1 GenAI integration across courses

2.2 Ethical Integration of GenAI in Courses

2.2.1 Course Redesign and Delivery Adaptations

One aspect of GenAI ethical integration was the redesign of courses and delivery methods. Here, students were not left to experiment on their own but were guided to test GenAI's capabilities, confront its biases, and validate its outputs through group discussions and diverse perspectives. They learned to treat AI as a supportive tool while sharpening their own analytical and literary skills. For example, in group projects, students had to use GenAI outputs but also challenge and validate them with their own research, cultural insights, and peer discussions. They marked what GenAI got right and wrong and then explained why. This way, GenAI became a tool for critical thinking, not a shortcut. Students also worked with both open and non-open access (copyright materials), learning when they could or could not use AI responsibly. The group activities, along with other synchronous and asynchronous tasks, are designed to integrate generative AI while ensuring that all topics reinforce the key concepts and frameworks outlined in the course module handbook. The organization of these activities not only allows students to experiment with technology but also ensures that the learning objectives are fully supported. Reorganizing course activities and including more challenging topics using structured safeguards enable students to engage with technology responsibly across various complex scenarios, without compromising the achievement of the intended learning outcomes. Through a series of experimentations, reflections, knowledge sharing with peers and the tutor, feedback analysis, and creative thinking, students can leverage their experiential learning for future explorations. This process helps them identify the potential applications of the technology as well as possible cases of misuse. These strategies are explained below.

Challenging Group Tasks with Detailed GenAI Usage Instructions: Challenging Classroom Group Tasks with Detailed GenAI Usage Instructions

A key element of course redesigning was the way group work was structured. For example, in the very first week of the management consulting course, students worked on a real-world case of an Indian restaurant chain exploring international expansion (Table 2.1) in the classroom. Each group had to research potential markets and rank them based on multiple factors. This requires students to bring together different kinds of evidence: GenAI-generated insights, peer knowledge from their own countries and experiences, and data from reliable sources such as journals and professional tools, followed by the justification of these reasonings during the presentations.

The design ensured that GenAI was not treated as an unquestioned authority. Instead, students were asked to highlight AI-generated content that proved accurate (marked in blue) and content that turned out to be incorrect (marked in red). The students are provided with the GenAI usage instructions (Table 2.2) to help them with how they could use GenAI and what limitations to watch for. In their presentations, they had to explain both—why some AI insights held up under scrutiny, and why others failed. This process did two things at once: it showed students that GenAI could be a tool to get directions about the research process and some market insights (which need to be validated and improved), but it also forced them to confront its biases, hallucinations, and inconsistencies. Through group discussions, cross-cultural perspectives, and peer validation, they developed stronger analytical and critical thinking skills.

Table 2.1 Group work of week 1 of management consulting course (M563)

• An Indian restaurant chain based in Mumbai has experienced significant success and popularity in the local market. With a reputation for authentic Indian cuisine and excellent customer service, the restaurant chain is now considering expanding its operations internationally. However, the company is unsure about which country to target for its first international venture.
• *Your group is the team of consultants specialized in food industry and market research. Do rigorous market research to identify potential international markets for expansion.*
• *The advice should include:*
– *Top three countries ranked for the different market selection factors*
• *Briefly mention the research methodology and data sources employed based on which the countries were ranked against multiple aspects (market selection factors).*
• *Your group also need to reflect on the information generated by the GenAI technology, considering the AI limitation (Table 2.2).*

The group work is designed in a way that students collaborate across diverse experiences and nationalities to strengthen their analytical and literary skills. By sourcing information from multiple references and validating it with the knowledge of group members—whether linked to their country of residence, nationality, or prior work experience—students reduce overreliance on GenAI outputs. They will further present their findings to at least two other groups, engage in discussions to reconcile conflicting results, and converge on a final set of values, country rankings, and market selection factors. The final work is presented in front of entire students for final validation of the solution, together with the discussion about technology's possible use cases (functional value) and limitations. This collaborative process directly addresses the following major limitations of GenAI[1] (Gupta, 2025):

- Biases *(systematic errors in AI models that reflect prejudices in training data, leading to unfair or skewed outcomes)*
- Instabilities *(variations in AI output due to minor changes in input data, affecting the consistency and reliability of decisions/outcomes)*
- Hallucinations *(instances where AI generates plausible but incorrect or nonsensical information)*
- Black-box issue *(lack of transparency in AI systems, making it difficult to understand or explain how decisions are made)*
- Ethical issues *(concerns related to fairness, privacy)*
- Unknowns *(unidentified issues or blind spots in AI models that could lead to unexpected outcomes)*

The GenAI usage instructions were provided to students to enable them to interact with the technology, explore its functional capabilities, and critically understand its inherent limitations (Table 2.2). These instructions became the structured guidance that helped students use GenAI responsibly to explore the complex tasks that require perfect integration of technological, literary, and critical thinking skills. For instance, they were told to try multiple prompts for the same query to test stability, cross-check all AI outputs with other sources, and take responsibility for final decisions rather than deferring to AI, and whether AI outcome is validated to be true or not, efforts are to be made to understand the logic used by technology to generate output. By asking students to explicitly map AI outputs to the six categories of GenAI limitations during their research work and the presentations, the course turned GenAI into both a practical tool and a critical object of study. The group work was based on the objective to help students solve the problem through gathering and analyzing information from multiple sources and to be able to critically challenge the GenAI outputs. By this way, students did not just learn with AI; they learned about AI, gaining insights into its strengths and limitations while sharpening their collaborative and ethical reasoning. This enables students to self-experiment with GenAI and identify potential use cases—as well as misuse cases—through responsible use grounded in reflective

[1] https://hbr.org/2024/05/ais-trust-problem

Table 2.2 GenAI instructions for the week 1 group work of management consulting course (M563)

- **You are free to use ChatGPT in an ethical manner. You have to do this:**
  - Gather information from the GenAI tool (using multiple prompts, including same prompt several times for same query); validate with information from other sources (journals, magazines, Google, other technologies like Google Market Finder and Google Ad preview), your own knowledge, group discussions with students of different experiences, and nationalities; and finally improvise the solution. This will help you to develop analytical and literary skills and explore technology functional values and limitations. Do not forget to highlight the information that came out to be correct and wrong using the GenAI technology *(see the below points).*
  - Mark the information (in presentation) that is taken from ChatGPT (but validated from insights from multiple data sources) in **blue color**.
  - Mark the information (in presentation) that is taken from ChatGPT (which was proved to be wrong based on the insights from multiple data sources) in **red color**.
  - Explain the blue- and red-colored information in detail during your presentation, clearly labeling each item according to the AI limitations listed below.
- **The types of issues to consider while using GenAI, along with how they should be detected, are outlined below:**
  - **Biases:** ChatGPT might recommend mostly English-speaking or Western countries for expansion (USA, UK, Canada), overlooking non-Western but high-potential markets (e.g., UAE, Singapore).
  - **Instabilities:** Running the same prompt twice could yield slightly different country rankings or contradictory justifications.
  - **Hallucinations:** ChatGPT might invent numbers, for instance about potential market in a country.
  - **Black-box issue:** You will not see the reasoning behind the technology outputs, for instance why technology rated one country higher than the another.
  - **Ethical issues:**
    - Do not input any personal information.
    - Do not upload any copyrighted documents or information.
    - Make independent decisions; do not let AI dictate outcomes.
    - Use AI respectfully; avoid harmful or manipulative outputs.
    - Ensure that all group members have equal access to AI tools. Help your group members with the technology through peer learning.
    - Verify AI outputs; take accountability for decisions.
    - Validate outputs with group members' cultural perspectives.
    - Avoid compromising others' privacy.
  - **Unknowns:** During your group discussions, you may uncover hidden or unforeseen limitations of AI that are not immediately apparent.

synthesis, that is, making informed decisions about when to engage AI and when to rely on one's own reasoning, based on critically integrating information from multiple sources, including generative AI. The GenAI usage instructions are tabulated in Table 2.2.

Challenging Group Tasks (Asynchronous Activities on Online Learning Platform)

The course project communication and stakeholder management (M562) was delivered on-campus as a face-to-face module, where students worked on group tasks and engaged in tutor-led discussions. To complement this face-to-face learning, the course was supported by a web-based learning system called Canvas. On Canvas, the learning journey was divided into nine weekly units, each aligned closely with the module handbook. These weekly units offered students structured content such as topic notes, case studies, links to online sources, and access to subscribed databases like EBSCO, quizzes, and discussion forums to extend classroom activities into asynchronous project-based work.

At the beginning of the module, students selected a potential project of their choice, which served as the catalyst for their weekly tasks on the Canvas. Each week, they revisited the classroom concepts through Canvas, where they replicated the types of group activities and tutor instructions completed in class as per the project task provided on the Canvas in the discussion forums. In weekly discussion forums, students shared their project-specific solutions as per assigned weekly tasks, drawing on classroom learning, research resources, and responsible GenAI use to develop evidence-based answers. Because these responses were visible to peers, the forums encouraged dialogue, critique, and feedback, which in turn helped refine ideas and strengthen solutions collaboratively.

The integration of GenAI was deliberate and structured. Students were guided by clear usage principles that encouraged them to experiment with technology responsibly while also reflecting critically on its strengths and limitations in the classrooms, for instance through challenging group works (*Section "Challenging Group Tasks with Detailed GenAI Usage Instructions: Challenging Classroom Group Tasks with Detailed GenAI Usage Instructions"*). This not only helped them deepen their subject knowledge but also taught them how to use GenAI as a knowledge collaborator. In their weekly activities and final submissions, students were encouraged to demonstrate how they had engaged with GenAI in their works through AI contribution form, GenAI integration rubrics, or AI contribution assessment form.

At the start of the module, each group selected a challenging professional project, chosen to reflect real-world complexity rather than simple scenarios. This project served as the anchor for all subsequent weekly tasks. The weekly asynchronous tasks on Canvas are briefly summarized below:

- In week 1, students introduced their project; described objectives, key stakeholders, and project characteristics; identified stakeholder types and roles; and reflected on stakeholder management across project phases, including the importance of negotiations and renegotiations.
- Week 2 focused on stakeholder motivation and expectations. Students examined how economic, social, and political factors influenced stakeholder involvement, evaluated their capabilities, and considered the implications of participation for fostering collaboration and addressing potential conflicts.
- Week 3 guided students in developing stakeholder management strategies and communication tactics. They presented a stakeholder management plan, explored engagement approaches, aligned communication strategies with project goals, referenced sample cases from the planning phase, and reflected on how communication needs might change across project phases.

- In week 4, students conducted stakeholder analysis using structured frameworks, prioritized stakeholders based on importance and influence, and reflected on insights gained from sample cases.
- Week 5 emphasized stakeholder engagement and relationship-building strategies. Students described specific actions to involve stakeholders in project activities and decision-making, referenced real-life business cases, and reflected on the role of engagement in long-term value creation and sustainable project outcomes.
- Week 6 focused on managing difficult stakeholders and resolving conflicts at multiple levels—downward, upward, and sideward/outward. Students drew lessons from business cases and reflected on how stakeholder management strategies impacted project progress and outcomes.
- Week 7 guided students to develop a comprehensive project communication plan, detailing communication types, methods, channels, reporting, and action plans. They reflected on factors influencing communication approaches, adapted strategies to stakeholder needs, and referenced sample cases demonstrating effective project communication.
- In week 8, students analyzed ethical issues in stakeholder management and project communication, considered mini-case dilemmas, and reflected on how ethical practices contribute to trust, integrity, and sustainable stakeholder relationships.
- Finally, week 9 consolidated learning by guiding students to create and implement a comprehensive stakeholder strategy, explore alternative approaches, and reflect on the potential long-term value created through effective stakeholder management.

The classroom learning—through examples, case studies, and group work—sets the baseline for the scenarios students must consider. In the weekly tasks, the aim is not simply to create a communication plan or stakeholder strategy for a hypothetical case. Instead, students must take the challenges and insights learned across the course, validate them through further research, narrow them to their selected project, and produce rational, evidence-based, and contextually valuable responses. While GenAI can play an assistive role in such tasks, when the problems are challenging and the requirements must be self-identified through classroom content, case studies, and group work, it cannot replace the critical thinking, judgment, and problem-solving that students apply to adapt theory to complex, real-world situations and arrive at optimal, contextually valuable solutions.

Over the 9 weeks, this design created a cumulative learning experience. Students tested assumptions, identified the boundaries of their project contexts, sought optimal approaches to communication and stakeholder management challenges, and compiled the weeks' outputs into a richer, practice-oriented portfolio for their chosen project. The whole design was deliberately aligned with the module handbook outcomes and aimed to blend theory, classroom practice, peer review, and GenAI-supported exploration into a cohesive learning journey by gaining rich, practice-oriented insights. Their chosen projects became the lens through which they applied the entire course learning, blending theory with real-world application in a way that was reflective, iterative, and deeply engaging.

Detailed, Multipage Files: Both Open Access (to Explore Technology Analysis Capabilities) and Copyrighted (to Examine Copyright Issues) in Group Works

Another key element of the course redesign was to build on the learning students gained through collaborative group work by increasing both the size and complexity of the documents used as data sources for secondary research, including open-access and non-open-access materials. The course activity was deliberately designed around large, multipage files, since such documents pose challenges for both human readers and GenAI systems. By working with these files, students tested AI's ability to manage scale and complexity while also confronting legal and ethical constraints. They experience how AI manages summarization, cross-referencing, and content extraction at scale—seeing where it succeeds and where it struggles. The instructions for the group work and nature of the group assignment remain the same as tabulated in Tables 2.1 and 2.2. In such group work, the students work with detailed, multipage files drawn from two categories:

- **Open-access materials:** These files are uploaded into GenAI systems for direct analysis, allowing students to explore how technology handles complex, lengthy content. When working with open-access documents in combination with GenAI and other data sources, the following occurred:
  - Students uploaded and processed these documents with GenAI.
  - They experienced how AI tools summarize, extract arguments, and cross-reference within long texts and the need to validate the information with data collected from multiple sources, including personal and professional experiences.
  - The information was also collected by involving multiple data sources, including manual analysis of the documents uploaded to the GenAI and group members' experiences to improve the GenAI outcomes to generate optimal and creative solutions to the problem.
  - They also saw the limitations of AI with large files, such as incomplete retrieval, hallucination, oversimplification, or failure to track nuanced arguments across a long text.
  - Based on the group work involving larger files, students grasped understanding of technology functional capabilities and limitations in handling large-scale information.
- **Copyrighted materials**: Although students can download these texts through university subscriptions, they are explicitly instructed not to upload them into GenAI systems to respect copyright laws. Instead, students read these documents manually and then use GenAI in an indirect way: by prompting the tool with targeted queries and combining the outputs with their own data gathering across multiple sources. When working with closed-access documents in combination with GenAI and other data sources, the following occurred:
  - Students had access to these through university subscriptions but were instructed not to upload them into AI tools to respect copyright.

- Instead, they manually read the texts and then used GenAI for indirect tasks, such as:
  Generating background knowledge or understanding of the concepts
  Testing interpretations with general queries
  Gathering additional evidence from other data sources to clarify their doubts and validate or improve the knowledge they acquired through manual reading
- The group members combined human critical reading with AI-supported research while staying within ethical and legal boundaries. This design ensures that students gain hands-on experience in responsible AI use, learning when and how they can integrate copyrighted information without violating ethical or legal boundaries, driven by their analytical and literary skills working in diverse groups.

ChatGPT is continuously evolving, with new models such as GPT-5 being launched that offer higher functionality and improvements compared to earlier versions like GPT-4.5 and GPT-4. Despite these advancements, concerns remain about the limitations of generative AI. It is clear that students should view technology as an assistive tool rather than a complete substitute. To validate and refine AI-generated outputs, students still need to rely on traditional methods of generating insights, such as analyzing multiple data sources. When working with larger files, students have observed that technology can efficiently summarize lengthy documents and answer basic questions. However, the output still requires verification, and only through manual document analysis, supported by human expertise and deductive reasoning, can meaningful connections between different pieces of information be established to answer complex and advanced queries.

Thus, in both types of groups' works, the students tested the functional capabilities and limitations of the technology. The reflection helped them to understand what AI can achieve when given full access to large open documents, what gaps appear when it cannot directly process a file (or it does), and how human literary skills and AI prompting can complement each other. As the group work progressed through the weeks, the skills and understanding gained helped them to identify possible use cases, scope, and limitations of the technology, along with the strategies to overcome them (by integrating traditional research methods with GenAI and avoiding overreliance on the technology). In either case, whether students worked with open-access documents that could be uploaded into GenAI systems or closed-access (copyrighted) documents that had to be read manually, the core learning task remained the same: data had to be gathered from multiple sources, critically examined, and then collectively analyzed in groups to reach a solution for which digital literacy, information literacy, and critical thinking and analytical skills are important. In both scenarios, students discovered that no single output from GenAI is sufficient. Cross-checking, triangulation, and group-based critical reflection were necessary to deal with limitations such as biases, instabilities, hallucinations, and ethical concerns. This reinforced the reality that responsible use of AI requires both technological proficiencies, for instance to find areas where technology could accurately work and reduce the manual efforts, and human judgment driven by experience and analysis of data across multiple reliable sources. The most reliable solutions emerge when AI is used in combination with collaborative, multisource data analysis.

Underexplored and Applied Research Topics Are Introduced in the Module Handbook

Final aspect of the course redesign and delivery adaptation was to strengthen the real-world connection of the course. To do this, the topics that were less covered in academic literature (underexplored) but highly applied in practice are covered as part of the course. For example, Table 2.3 shows two important areas covered in the management consulting course owing to the real participation in these areas of the tutor. The students were introduced to the real consulting case of the Spanish startup that globalized with its sanitization product and how the students could replicate the same study for the same startup with its more innovative product and changed market situations. Another key area introduced was based on ongoing consulting with the libraries, including academic and public libraries, looking to start or improve their business support services. Students were provided with the case of some libraries and asked how they will research the new ways of improving business support services for some of the libraries and how their consulting reports will look like. These topics encouraged students to go beyond existing research and combine GenAI with their own creativity and traditional methods. By doing so, they saw that while AI could provide starting points, original thinking and evidence-based reasoning remained central to academic and professional work owing to the open-ended nature of the problem, unique working context of the startups and libraries, and continuously evolving business environments. Table 2.3 tabulates these two topics in management consulting course.

2.2.2 Assessment Brief Redesign

The assessment brief was redesigned to integrate a clear AI policy for the course, providing students with guidance on the ethical and responsible use of generative AI tools as a knowledge collaborator with no substantial contribution. The assessment tasks were reframed as applied research activities built around open-ended, real-world problems, encouraging students to apply theory critically and creatively. To support this, an AI Contribution Assessment Form was introduced. This later evolved into a GenAI Integration Rubric, which outlines potential uses and misuses of AI in relation to various research tasks. This rubric was explicitly mapped against the roles defined in the CRediT

Table 2.3 Underexplored and more applied research topics introduced in the module handbook of management consulting course (M563)

Week 8: Real consulting case: Globalization of Spanish startup in pandemic (including hands-on with technologies)
Week 9: Real consulting case: Exploring innovative library entrepreneurial services for business support

taxonomy, helping students to understand how AI could responsibly contribute across different stages of the research process while maintaining academic integrity. The same form could be adapted by the students to highlight for which tasks they used the tool, and finally reporting the level of the usage. An assessment brief is a document that outlines the requirements and expectations of a task to be assessed as part of the course evaluation. It describes what students are expected to do, the scope of the work, the assessment criteria, and the conditions for submission, thereby providing clarity and transparency in the assessment process. As part of the assessment brief redesign, the tasks were redefined as Applied Research Assignments with open-ended problems. The assessment rubrics were redesigned to align with the applied nature of the tasks, and a Special Instructions section was added to explain the course AI policy, including permission to use Generative AI ethically and the requirement to submit an AI Contribution Form with the final assignment. The AI Contribution Form includes a reference to either the GenAI Integration Rubric or the AI Contribution Assessment Form, depending on which instrument the course module leader chooses to apply. The assessment brief, the AI Contribution Form, and either the GenAI Integration Rubric or the AI Contribution Assessment Form are uploaded as separate documents on the university's web-based learning system, Canvas. These aspects are explained in detail below.

Applied Research Assignments with Open-Ended Problems

As part of the redesign of assessment briefs, one solution was to redesign the assignments in a way that GenAI alone could not provide complete solutions. The assignments were structured in a way that students had to engage in applied, research-based tasks that demanded creativity, critical thinking, and justification of their choices, leveraging the experiences they have gained with the applied nature of group tasks throughout the course.

Assignments were restructured into open-ended consulting problems that mirrored the kind of work students might face in their professional careers. For instance, the management consulting course assignment asked students to act as consultants hired to improve a university's startup services, requiring them to analyze, propose strategies, and design implementation plans (Table 2.4). Their task was to design a strategic consulting proposal for improving a university's startup services. This was not a simple academic exercise; it required them to analyze the current state of entrepreneurial support at a chosen university, identify strengths and weaknesses, and then develop a set of strategic recommendations. Students had to move beyond theory by building an actionable implementation plan, complete with timelines, resources, and stakeholder strategies. The challenge of this assignment was that no GenAI tool could generate a finished solution: while AI could suggest possible directions or highlight common startup services, students had to contextualize those ideas, filter them through evidence, and align them with best practices in consulting. This design encouraged students to see GenAI as an initial brainstorming partner, but it was their critical thinking and application of research that shaped the final outcome.

The same approach could be applied not only to the assignment of final grades but also to intermediate assessments throughout the course, including ungraded ones. Using a

Table 2.4 Applied research assignments with open-ended problems for management consulting course (M563)

Title: Strategic consulting proposal for enhancing university startup services

Background: University startup services are integral in fostering entrepreneurial growth by providing support in various areas such as raising awareness, developing entrepreneurial skills, obtaining market insights, and facilitating commercialization efforts. As management consultants, your team's objective is to propose strategic recommendations to optimize the effectiveness and impact of university startup services in achieving these goals.

Tasks: Take any university of your choice that is actively offering support to the entrepreneurs. Assume that you are a consultant hired by the university to improve their entrepreneurial services. Do rigorous research, and then prepare the **consulting report (format attached below).**

The objective of the consulting project is to:

- Analyze the current state of university startup services, considering their effectiveness in supporting entrepreneurship and identifying areas for improvement *(awareness-raising strategies, entrepreneurial skill development programs, market insight acquisition initiatives, and commercialization support mechanisms).*
- Develop strategic recommendations to enhance the impact and reach of university startup services, aligning with best practices in management consulting.
- Create an actionable implementation plan outlining key initiative, timelines, resource allocation, and stakeholder engagement strategies.
- Present the consulting proposal in a compelling manner, demonstrating a deep understanding of the challenges and opportunities in the university startup ecosystem.

Consulting project report

- **Executive summary**
- **Introduction to university startup services and their importance** (*mention here name of the university along with weblink)*
- **Analysis of current state**
  - SWOT analysis of university startup services
  - Identification of key challenges and opportunities
- **Strategic recommendations**
  - Awareness-raising strategies
  - Entrepreneurial skill development programs
  - Market insight acquisition initiatives
  - Commercialization support mechanisms
- **Implementation plan**
  - Timeline for execution
  - Resource allocation and budgeting
  - Stakeholder engagement strategies
- **Evaluation metrics for success**
- **Conclusion and reflection**

Note: Remember that your solution should be based on evidence to make your recommendations more rational, practical, and feasible.

variety of assignment formats—such as research papers, podcasts, and videos—helps motivate students to focus on creative solutions and translate their research into innovative outputs. The excitement of creating and sharing these unique submissions encourages originality while discouraging unethical use of technology. For example, submitting a research paper can motivate students to incorporate tutor feedback, refine their work, and even consider submitting it for publication, supporting their career goals, particularly for those aiming for higher education or competitive job opportunities. This approach promotes practical and applied research and reduces their stress associated with submitting long, detailed assignments. In reality, students get exposed to many media for promoting and sharing their unique insights addressing the real problem. Their engagement with real-world problems helps ease the pressure of heavy academic workloads and tight deadlines, which are factors linked to students' use of ChatGPT (Abbas et al., 2024).

The approach of using applied, research-based assignments with unique reporting formats, apart from the courses reported in this book, was also observed in a project at MrciS. One such experience had been briefly reported by Gupta and Gupta (2024). According to Gupta and Gupta (2024), one student group was tasked with creating a communication plan for a company facing global stakeholder engagement challenges. The group reported that, while not substantial, generative AI made important contributions across five roles: conceptualization, investigation, methodology, writing (original draft), and writing (review and editing). The group used ChatGPT to draft a tailored plan, refining it with expert input, secondary sources, and stakeholder interviews, relying on the technology in an assistive role rather than as a substitute. AI-generated data and other data collected from other sources were refined iteratively, and the final plan was submitted to the startup with faculty feedback. This assignment increased motivation, reduced stress from long submissions, enhanced team participation, and promoted knowledge sharing. Close collaboration between faculty and the startup helped students understand ChatGPT's capabilities and limitations, refine research skills, develop creative solutions, disseminate insights interactively as a short report, and create the short video summarizing key findings.

In another course, project communication and stakeholder management, the applied assignment took the form of a group video simulation (Table 2.5). Students were asked to replicate a real project environment, taking on stakeholder roles such as project manager, client, investor, or community representative. Their task was to research an organizational problem, identify relevant stakeholders, and then create a video assignment.

The video not only demonstrates a real project environment but also clearly demonstrates a well-researched problem, a rational solution addressing a stakeholder management or communication challenge, and reflections that could provide useful lessons for peer organizations. Here again, GenAI could play a limited but supportive role, for instance, helping generate role-play scenarios, suggesting communication frameworks, or providing ideas for stakeholder analysis. However, true learning came from students' ability to tailor these suggestions, improvise realistic interactions, and present an engaging, research-informed solution for the unique problem and any aspect pertaining to the course they have selected to be researched for optimal solution. This assignment blended creativity with evidence-based reasoning, ensuring that students demonstrated ownership of their work while still exploring how AI could support, but not replace, their learning.

In the Special Instructions of the Assessment Brief (Table 2.6), the AI policy is explicitly stated, together with the requirement to submit the AI contribution form, along with either the GenAI integration rubric form or the AI contribution assessment form (Table 2.8). The special instructions are tabulated in Table 2.6.

Table 2.5 Applied research assignments with open-ended problems for project communication and stakeholder management course (M562)

Group video: Simulating stakeholder management and communication in a project environment **Objective** To demonstrate an understanding of stakeholder management and communication concepts by creating a group video that simulates a typical project environment and effectively illustrates stakeholder engagement and communication strategies driven by your researched solutions. **Instructions** • Create a group of maximum four students. • It is strongly recommended that you pick an organization or company you are familiar with and select a professional project. For instance, you can think about any existing product and its limitations. These limitations could be an interesting project for your company to work on (for instance, improved product as the outcome or discussions among project team about possible integration of GenAI for product development activities). *The project should be a professional project.* This project will be the focus of the video. • The project could belong to any industry or sector (e.g., construction, marketing, healthcare, technology). • Your group should select a professional project problem, research the organization and its issues, and identify relevant stakeholders. Within each group, assign specific roles to each member based on the stakeholder types relevant to the selected project. Examples of stakeholder roles include project manager, team member, client representative, investor, regulatory authority, community representative, etc. *(depending on the problem you have considered through rigorous research)*. Each group should include a diverse mix of stakeholder roles. • Conduct research and discussions to understand the characteristics, interests, motivations, and potential concerns of their assigned stakeholder roles. Consider the power, influence, and level of involvement each stakeholder may have in the project. This should then be considered for the problem you selected for the final assignment. • Choose a challenging aspect of the project (e.g., engagement, communication, coordination), and develop a solution addressing this stakeholder management challenge. The group video should simulate the project environment, showing stakeholder interactions, challenges, and how your proposed solution addresses the problem. • References used for research should be displayed at the end of the video. • You should also reflect on your experiences with stakeholder management and communication (aligned with your selected project) that could become useful lessons for other organizations. • Each group member should briefly identify himself or herself (particularly by name) at the beginning of the video. **Note:** (**a**) Upload the video along with the ***AI contribution form and GenAI integration rubric form (refer to special instruction section of assessment brief, classroom discussions, AI contribution form, and GenAI integration rubric form for more details)***. (**b**) This assignment is research based. Your group should not only simulate a real project environment but also clearly demonstrate a well-researched problem, a rational solution addressing a stakeholder management or communication challenge, and reflections that could provide useful lessons for peer organizations.

GenAI Integration Rubric

The GenAI integration rubric gave students a structured way to become aware and reflect on how, where, and to what extent GenAI could be used in their assignments. The GenAI integration rubric was not designed as a rigid checklist but as a flexible guide to help

Table 2.6 Special instructions about GenAI policy in assessment brief

You are free to use the generative artificial intelligence tools like ChatGPT. However, you must use them to enhance your learning outcomes rather than as a replacement for the manual learning efforts. This means that such tools can be used as a "**knowledge collaborator**" in an ethical manner. Using technology as a knowledge collaborator means drawing on it for guidance or support without letting it make a substantial contribution. Ethically, AI must never be considered a co-author of your work. You must understand that sometimes these tools generate wrong information. Any information obtained from such tools should be considered as merely "hypothesis." You must employ your creative problem-solving skills to improve the information generated by such tools. Your assignment will be evaluated based on its creative content and presentation, which should be driven by your creativity and problem-solving skills. Please refer to the AI contribution form and the instructions provided during your course sessions. **AI contribution form** should be used to report the AI tool contribution. The AI contribution form is designed to help you understand the GenAI policy adopted for the course and to provide a transparent record of how you have used and acknowledged AI tools (including cases where you did not directly use AI-generated output in your assignment). Specifically, the form requires you to report the following information: • Screenshot of the prompts given to ChatGPT. • Which information in your assignment is provided by the AI technology (map it with the prompt information), and how much it was adapted to meet your needs or improve its accuracy? **Note:** Please keep in mind that failing to disclose the AI contribution or obscuring the actual contribution will result in an accusation of academic misconduct.

students recognize where technology might be meaningfully applied in their research-based assignments.

Drawing on the CRediT taxonomy, the rubric mapped out possible roles that GenAI could play across the research process—such as supporting conceptualization, assisting with investigation, improving clarity in writing, or asking the technology for feedback about the assignment.

For each role, the rubric includes guiding questions tailored to the role-specific use of generative AI (GenAI). For every question, the tutor provides illustrative examples aligned with the final assignment (as outlined in the assessment brief), showing both how GenAI could appropriately support the role and where its use might constitute academic misconduct. For instance, under the investigation role, one guiding question is: *Did you conduct a literature survey to inform the investigation?* An appropriate use of GenAI in this case could be experimenting with it to summarize key findings from multiple academic papers to help identify gaps in existing research. Conversely, an example of academic misconduct would be relying solely on GenAI-generated summaries without consulting and verifying the original sources, which could result in misrepresentation or omission of critical information. For each example, across every question and role, the rubric also specifies whether acknowledgement of GenAI use is required. Finally, a rating column enables students to self-assess the level of AI involvement in their work—low (less than 20%), medium (20–50%), or high (more than 50%)—thereby ensuring transparency and adherence to academic integrity (Table 2.7).

Table 2.7 GenAI integration rubric for management consulting course (M563) (*adapted from Gupta and Gupta* (2024) *and Gupta and Nyamapfene* (2025), *reused with the permission of the publisher; the figure remains under the standard IEEE license*)

AI contribution role	Questions	Example activities that could benefit from GenAI	Example activities that could result in misconduct (*applicable for category 2*)	AI acknowledgement required? *Specify in which part (A or B) of the AI contribution form the acknowledgement is made*	Rating Low *(less than 20%)* Medium *(20 to 50%)* High *(above 50%)*
Conceptualization	*Did you use GenAI to define the research problem or objectives?*	Drafting problems or objectives that highlight gaps across four areas, i.e., raising awareness, developing entrepreneurial skills, obtaining market insights, and facilitating commercialization efforts (more generic). This is to get more familiar with generic problems	Copying generic AI-generated problems or objectives without tailoring to the university's context, for which you need rigorous secondary research of universities	Yes	Low, medium, high
	Did you use GenAI to identify research questions or hypotheses?	Generating research questions based on research gaps identified through GenAI *(if used)* and own research	Submitting AI-generated questions without refinement and without considering research-supported insights about specific university	Yes	Low, medium, high
	Did you use GenAI to develop the initial idea or concept for the project?	Brainstorming project ideas such as campaigns, training modules, market studies, or commercialization practices of universities	Allowing AI to define the project concept without critical thinking and without researching university context	Yes	Low, medium, high
Data curation	*Did you use GenAI for data management tasks, for instance, organizing, cleaning, describing, enhancing, storing, and preserving data?*	Get suggestions for data sources, creating data definitions, and databases with already collected data (if available)	Relying on AI-generated output as part of data management without manual validation or provenance checks	Yes	Low, medium, high

Formal analysis	*Did you use GenAI to apply statistical or computational techniques to analyze data?*	The collected data could be analyzed, for instance categorized into themes to cross-check those generated by team	Submitting AI-organized data without checks and without rigorous analysis	Yes	Low, medium, high
	Did you use GenAI to identify the patterns, trends, or relationships in the data?	The collected data could be analyzed, for instance coded and converted into themes (in case of grounded theory qualitative data analysis), to be cross verified with the ones created manually or using qualitative data analysis tools	Treating AI-identified patterns as validated findings without verifications	Yes	Low, medium, high
	Did you use GenAI to synthesize the data collected from multiple sources into meaningful insights?	The collected data could be analyzed, and meaningful insights be generated *(which will be broader)*. This could help in understanding if manual insights generated using your creative thinking miss some perspectives or directions and hence warrant further investigation	Presenting AI-synthesized conclusions without verifying source consistency or citing originals, or without using your creative thinking skills. The insights leading to meaningful discussions should be rational and fully aligned to the selected university's real context driven by scientific research	Yes	Low, medium, high

(continued)

Table 2.7 (continued)

AI contribution role	Questions	Example activities that could benefit from GenAI	Example activities that could result in misconduct *(applicable for category 2)*	AI acknowledgement required? *Specify in which part (A or B) of the AI contribution form the acknowledgement is made*	Rating Low *(less than 20%)* Medium *(20 to 50%)* High *(above 50%)*
Investigation	*Did you use GenAI to perform data collection either primary or secondary?*	Use AI to locate secondary sources or datasets and case studies relevant to selected university	Publishing or using fabricated or AI-simulated "collected" data as if gathered from real respondents (primary data) or from secondary sources	Yes	Low, medium, high
	Did you use GenAI to do a literature survey to inform the investigation?	Ask AI to locate themes, summarize papers (open access), and identify gaps as applicable generically for universities (then verify sources) or specific universities (if relevant research studies are available). This will be refined further through scientific research	Relying on AI summaries and hallucinated citations without checking primary sources and without manual analysis of the research papers	Yes	Low, medium, high
	Did you use GenAI to conduct experiments, surveys, or observations?	Use AI to draft experiment/survey/ case study designs, sampling approaches and observation checklists for the four focus areas. This should be refined further through research	Using AI-generated information directly without further verification and modifications based on research insights	Yes	Low, medium, high

Methodology	*Did you use GenAI to design the overall approach or methodology for the project?*	Considering the selected university context, time frame for getting ethical board approvals for primary research, and resources that team has, GenAI could propose a research methodology for the project	Copying an AI-proposed methodology wholesale without aligning to assignment constraints or ethics	Yes	Low, medium, high
	Did you use GenAI to select appropriate research methods and techniques?	Considering the selected university context, time frame for getting ethical board approvals for primary research, and resources that team has, GenAI could propose some suitable research methods that could be applied. This needs to be further researched to assess suitability for the project	Accepting AI's method choice without justifying feasibility, ethics, or alignment with case context	Yes	Low, medium, high
	Did you use GenAI to design research study protocols and procedures for data collection and analysis?	Draft survey instruments based on indicators drawn on your research findings and interview guide for qualitative interviews Use AI to draft informed consent text, data collection, and analysis procedures to be applied	Using AI-generated survey instruments or interview guides, research protocols without modification (not grounded on realities as revealed through research process)	Yes	Low, medium, high

(continued)

Table 2.7 (continued)

AI contribution role	Questions	Example activities that could benefit from GenAI	Example activities that could result in misconduct *(applicable for category 2)*	AI acknowledgement required? *Specify in which part (A or B) of the AI contribution form the acknowledgement is made*	Rating Low *(less than 20%)* Medium *(20 to 50%)* High *(above 50%)*
Software	*Did you use GenAI to develop any software application that was used in any research activity, for instance, data analysis, modeling, or visualization?*	Not applicable for the assignment			Low, medium, high
	Did you use GenAI to implement the algorithms or working software for specific tasks?	Not applicable for the assignment			Low, medium, high
	Did you use GenAI to test and debug software components that are the outcome of the research or those required by the research?	Not applicable for the assignment			Low, medium, high

Validation	*Did you use GenAI to check the accuracy and reliability of research findings, for instance by performing research replications?*	Prompt AI to suggest replication checks or alternative explanations for findings. Treat them as hypothesis triggering further research	Treating AI's replication suggestions as definitive proof without independently rerunning checks and getting more perspectives through further research	Yes	Low, medium, high
	Did you use GenAI to verify the validity of assumptions and methodologies used in the project?	Use AI to surface hidden assumptions, robustness checks, and alternative methodological choices	Accepting AI's verification as substitute for methodological review or ignoring contradictory evidence	Yes	Low, medium, high
Visualization	*Did you use GenAI to create or redesign the visual representations of data or research findings?*	Generate Gantt charts or visuals of your research findings based on the data input by the team	Submitting AI visuals based on fictitious or AI-generated data	Yes	Low, medium, high
Writing—original draft	*Did you use GenAI to draft full or portions of the version of research papers, reports, or other project documents?*	GenAI can be used to improve expressions or statements of the assignment	Using GenAI text that does not align with the meaning of original text	Yes	Low, medium, high

(continued)

Table 2.7 (continued)

AI contribution role	Questions	Example activities that could benefit from GenAI	Example activities that could result in misconduct *(applicable for category 2)*	AI acknowledgement required? *Specify in which part (A or B) of the AI contribution form the acknowledgement is made*	Rating Low *(less than 20%)* Medium *(20 to 50%)* High *(above 50%)*
Writing—review and editing	*Did you use GenAI to review the research documents for clarity, coherence, and accuracy, for instance, improving expressions?*	Use AI for proofreading, clarity edits, and suggesting more concise ways to communicate findings and recommendations across the four areas	Allowing AI to rewrite major analytical content that is not grounded on your research and is not aligned with the original meaning of the text	Yes	Low, medium, high
	Did you use GenAI to edit the document for grammar, spelling, and punctuation errors?	Use AI to correct language issues, formatting suggestions, and citation style, so the final report reads professionally	Using AI editing to hide that substantive content was AI-generated rather than student-created	Yes	Low, medium, high
	Did you use GenAI for feedback and suggestions for improvement to enhance the quality of the writing?	Solicit AI feedback about document quality and then investigate further, incorporate improvements, and critically reflect	Incorporating AI feedback as independently validated improvements without further investigation and masking student's actual research work, for instance changing the actual research findings because of inappropriate use of the terminologies and inaccurate suggestions	Yes	Low, medium, high

The examples where technology could be used are not intended to be directly replicated in the assignment. Rather, they highlight specific aspects where GenAI might be experimented with, encouraging students to explore its potential while still integrating perspectives from other sources and not relying exclusively on GenAI. Examples of academic misconduct, on the other hand, are designed to raise awareness of situations where the use of technology in the manner described would be considered unethical. Although it is not possible to specify every possible case of appropriate or inappropriate use, these examples provide assignment-specific illustrations that help students recognize opportunities for responsible experimentation and identify practices to avoid. The rubric is flexible: new guiding questions can be introduced for each role, and multiple examples may be added at the intersection of each role and its corresponding questions. Students' experiences with technology throughout the course, for instance, when working in groups, will be instrumental in helping them better understand the rubric and complete it accurately for the final submission, alongside the AI contribution form and the final assignment.

GenAI integration rubrics served two purposes. First, it acted as a practical help guide, raising awareness of the types of activities where GenAI might be useful and where it should be avoided. Second, it became a reporting tool, requiring students to declare whether GenAI had been involved in their work and to what degree. To reinforce accountability, the rubric had to be submitted together with the assignment, alongside the AI contribution form, with clear indications of the contribution level (low, medium, or high). The contribution level percentages can be adjusted according to existing academic regulations or based on the tutor's assessment specific to the course. Importantly, the framework emphasized that not every taxonomy role would apply to every assignment and that in all cases GenAI's contribution should remain non-substantial, supporting but never replacing the student's own intellectual effort.

Through this dual role—as guidance and as documentation—the rubric and form together ensured that students gained both practical support in using the technology and ethical awareness about acknowledging its role. In this way, they learned that transparency and critical judgment are as important as technical skills when integrating emerging tools into academic work. Table 2.7 shows the GenAI integration rubrics for management consulting course assignment.

Students will not be using GenAI across all CRediT roles or research activities, so the GenAI rubrics for actual assignments will be less populated. Acknowledgement of GenAI support is required for any role or activity, including brainstorming, although in some universities, such minor tasks are not normally acknowledged. The acknowledgement will take two forms: one where students detail the reasoning behind using GenAI along with the prompts employed *(Part A of the AI contribution form)*, and another where they provide a broader summary of the activities for which GenAI was used *(Part B of the AI contribution form)*. The former refers to activities where GenAI outputs were directly used in the assignment, either fully or partially, and/or helped reshape the research work. The latter refers to activities where GenAI did not reshape the research and/or its outputs were not directly used in the assignment. For example, using GenAI to gain initial ideas about the project should be acknowledged in Part B of the AI contribution form, without providing details of the interactions. Activities such as using GenAI for grounded theory to refine themes generated manually by the team, and subsequently improving those themes through further research, should be acknowledged in Part A. If the themes identified by GenAI align with those already identified through research or are validated as incorrect, and the team decides not to include any GenAI-generated themes, with no new themes emerging from its recommendations, then the research work is not reshaped and the GenAI output is not directly used. In this case, it should be reported in Part B of the AI contribution form. In either case, any use of GenAI must be reported in the AI contribution form, and students must ensure that its contribution remains non-substantial. For assignments that are more practical and open-ended, GenAI contributions are expected to be minimal, and the acknowledgements made in the AI contribution form, as per the course policy, will be limited.

AI Course Policy

From the very beginning, the courses were framed by a clear AI course policy, shared with students well before the course commenced. The AI course policy clearly mentions if GenAI use in the course is allowed or not. If allowed, then this policy supports the ethical use of technology in a way that GenAI neither poses a threat to academic integrity nor serves as a shortcut to bypass learning, but as a knowledge collaborator with no substantial contributions. Students were encouraged to use the technology to support their work—for brainstorming, refining language, or exploring perspectives—but always with the understanding that intellectual responsibility and ensuring non-substantial contribution of the technology in the assignments rested firmly with them. The policy was communicated through the AI contribution form, which was uploaded online together with the assessment brief and the GenAI integration rubric well before the course began. By making these documents available from the start, students were given clear, transparent guidance on how GenAI could and could not be used in their assignments.

The contribution form served a dual function. First, it worked as part of the AI policy framework, providing practical instructions; for instance, students had to log their AI interactions, including the prompts they used, the outputs received, how they adapted or corrected those outputs, and their reflections on the value and limitations of the technology. This encouraged students to see AI-generated material as provisional hypotheses that required validation, rather than as ready-made answers.

Second, the form also became an integral tool for course evaluation. At the point of submitting their assignments, students were required to submit the completed contribution form alongside their work. This not only ensured transparency but also provided structured data on how students were engaging and reporting on those interactions with the technology. By reviewing these forms, instructors could monitor patterns of GenAI use, identify common strengths and pitfalls, and ensure that technology usage is non-substantial and academic integrity is maintained. This form serves to acknowledge GenAI use, whether it directly shaped the research and/or its outputs were used fully or partially in the assignment, as well as cases where it did not reshape the research and/or contributed outputs directly.

This form, together with more practical group works and continuous ethical usage of the technology, helped them to find possible use cases of the technology, which will vary across students owing to diversity of selected problems, cases, data sources involved, and proposed solutions. This further made students aware that technology itself will not be able to help them solve their final assignment, and they are expected to disclose the technology used through AI contribution form and GenAI integration rubrics. In this way, the form became both a safeguard for academic integrity and a learning tool that promoted ethical, transparent, and reflective engagement with GenAI.

By integrating the AI policy into the AI contribution form and linking it with the GenAI integration forms, students entered the course with clear expectations about *how* they could responsibly experiment with GenAI, *what boundaries they needed to respect*, and *how their use of the technology would be evaluated.* This not only created transparency but also reassured students that they would be supported in learning to use emerging technologies ethically, rather than penalized for exploring them. The AI policy informs students of the rules regarding GenAI use in the course. The GenAI integration form allows them to demonstrate how they applied these rules in practice and also serves possibly as usage guidance by providing suggestions for possible activities where technology could be used or not for the assignment and the need for proper acknowledgement. The AI contribution form, in turn, is used for reporting what they actually did with GenAI in their final assignments, including reflections on all reported interactions. GenAI integration rubrics map these interactions according to CRediT taxonomy roles while also capturing acknowledgement and students' self-assessment of the level of technology contribution. AI contribution form is tabulated in Table 2.8.

2.2.3 Course Evaluation Adaptations

Course evaluation has been adapted to ensure that students are assessed on the practical assignment based on their creativity, scientific rigor of their research, and their development of practical solutions to the selected problem, thereby eliminating the possibility of technology performing the work on their behalf. Students may use technology for non-substantial contributions, such as obtaining guidance or improving language, but such usage must be transparently reported through the AI contribution form and generative AI integration rubrics. By adopting this approach, students can transparently report their use of technology without concern that non-substantial contributions, if detected by generative AI detectors, would be treated as academic misconduct. Tutors evaluate the assignment based on the assessment rubrics, and any non-substantial contributions flagged by detectors can be assessed in context by reviewing the AI contribution form and generative AI integration rubrics, ensuring fair and informed evaluation. Evaluation is conducted according to the assessment rubrics outlined in the assignment brief, emphasizing practical application, alignment with learning outcomes, and effective, responsible integration of tools, including AI, into the assignment work.

Table 2.8 AI contribution form for management consulting course (M563) (with GenAI integration rubrics instructions)

AI contribution form
Based on the assessment brief of the course "***Management Consulting (M563)***," you were granted permission to use the generative artificial intelligence tools like ChatGPT. However, it is essential to employ these tools to enhance learning outcomes rather than as a replacement for the manual learning efforts. In this context, these tools are meant to serve as a "***knowledge collaborator***" and should be utilized in an ethical manner. The use of technology as a knowledge collaborator means employing it to obtain guidance, directions, or support during research or assignments without allowing it to make a substantial contribution to your final work. Ethically, the AI should not be considered a co-author of your submission (*and should not be eligible for it*). To ensure proper use, you can apply the GenAI integration rubric, which maps potential AI involvement across the CRediT taxonomy roles. This rubric helps you distinguish between activities where AI use is appropriate and those that could constitute academic misconduct. Note that: • Not all taxonomy roles may apply to your work. • Not all possible activities are listed in the rubric. Classroom experiences with GenAI, research-based exercises, and discussions about your assignments will help you apply the rubric effectively. Submit the filled GenAI integration rubric together with the AI contribution form and your final assignment. It is important to acknowledge that sometimes these tools generate wrong information. Any information obtained from such tools should be considered as merely "hypothesis." You must employ your creative problem-solving skills to improve the information generated by such tools. Your assignment will be evaluated based on its creative content and presentation, which should be driven by your creativity and problem-solving skills. Use the AI contribution form template (***enclosed below for your reference***) to report the AI tool contribution. In particular, the AI contribution form will highlight AI tool contributions by focusing on the following aspects (provide interaction details for each CRediT taxonomy role played by technology in your assignment): • Screenshot of the prompts given to ChatGPT. • Which information in your research paper is provided by the AI technology (map it with the prompt information), and how much it was adapted to meet your needs or improve its accuracy? Please keep in mind that failing to disclose the AI contribution or obscuring the actual contribution will result in an accusation of academic misconduct. We are committed to upholding academic integrity while also harnessing the full potential of technology to maximize our learning experience. You might not need to include the specific interactions in the AI contribution form if you only utilized an AI tool as a reference and did not directly use the information in your assignment. Nevertheless, it is still a good idea to express appreciation for the tool and briefly discuss how you used it in the assignment in the acknowledgement area.

(continued)

Table 2.8 (continued)

Part A (AI-generated information was directly used and/or helped reshape the research)
The technology use involved multiple interactions, with each interaction comprising several prompts to elicit specific responses from ChatGPT, enabling a comprehensive exploration of the topic that shaped and informed the research work validated, and/or refined GenAI outputs were subsequently incorporated, either fully or partially, into the research work and the final assignment. **Student names:** [*Write here names of all students of your group*] **Project title:** [*Write here title of your project*] **Interaction 1** **Date and time:** [*Mention here the date and time of interaction 1*] **Technology role:** [This role needs to be aligned with CRediT taxonomy and AI contribution assessment form] **Context:** [*Briefly describe why you used AI for this interaction*] **Purpose:** [*Explain the specific task or purpose for using AI*] **AI output:** [*Copy and paste or summarize the AI-generated content. Include all the outputs pertaining to the interaction*] **Student's contribution:** [*Describe how you integrated the AI-generated content into your project or analysis*] **Reflection:** [*Reflect on the value, limitations, or insights gained from this interaction. You may also focus on how the information presented by you builds on the one generated by AI tool*] **Interaction 2 (and onwards)** **Date and time:** [*Mention here the date and time of interaction 2*] **Technology role:** [This role needs to be aligned with CRediT taxonomy and AI contribution assessment form] **Context:** [*Briefly describe why you used AI for this interaction*] **Purpose:** [*Explain the specific task or purpose for using AI*] **AI output:** [*Copy and paste or summarize the AI-generated content. Include all the outputs pertaining to the interaction*] **Student's contribution:** [*Describe how you integrated the AI-generated content into your project or analysis*] **Reflection:** [*Reflect on the value, limitations, or insights gained from this interaction. You may also focus on how the information presented by you builds on the one generated by AI tool*] **Note: Include here all the interactions that resulted in the AI-generated information as directly used in the assignment and/or shaped the research work.** We would like to acknowledge the use of ChatGPT as an important source of information when conducting the study for this assignment. The data produced by the application as directly used in the assignment and/or how it reshaped the research work is mentioned in Part A. This was a useful resource that helped the team to gain a deeper grasp of the topic.

(continued)

Table 2.8 (continued)

Part B (*AI-generated information was not included in the assignment and/or did not significantly shape the research work*)
Acknowledgement We would like to acknowledge the use of ChatGPT as a valuable informational resource during the preparation of this assignment. Although the outputs from the application were not directly included in the final work and had limited influence on shaping the research, they provided useful initial insights that helped the team better understand the research topic. The tool was used for a variety of things as outlined below. ***(Explain how the tool assisted you here, aligned with CRediT taxonomy and GenAI integration rubrics using the following structure).*** *I acknowledge the use of ChatGPT (version XXXX) in the following roles and activities, which are also outlined in GenAI integration rubrics; the outputs were not directly used in the assignment and did not reshape the research work:* ***Conceptualization:*** *[For example, assisted in generating initial ideas for research themes].* ***Formal analysis:*** *[For example, cross-checked themes generated by the team using grounded theory with GenAI-generated themes; further investigation led to the decision not to incorporate the AI suggestions].* *Add as many as roles that apply.*

AI Contribution Form

As discussed in the previous section, the AI contribution form served a dual purpose. It guided students in logging their AI interactions—including prompts, outputs, adaptations, and reflections—encouraging them to treat AI-generated material as provisional insights rather than ready-made answers. The AI contribution form also serves to document the AI course policy, specifying whether GenAI use is permitted, and if so, how it should be used, what limitations to consider, and how students should record their technology use and reflections. Guidance on usage and limitations is linked back to the instructions that will be provided throughout the course, including specific examples such as the GenAI instructions for the group work (Table 2.2). At submission, the completed form accompanied their assignments, providing transparency and structured data on technology usage.

The AI contribution form, as presented in Table 2.8, is intended for courses where students are provided with the GenAI integration rubric (Table 2.7) and are required to complete and submit it alongside the final assignment. If the tutor chooses to use the AI contribution assessment form (Table 2.10) instead of the GenAI integration rubric, the AI contribution form (Table 2.9) should be used. The AI contribution form, as shown in Table 2.9, replaces references to the GenAI integration rubric with guidance specific to the AI contribution assessment forms.

The connection between GenAI integration rubrics and AI contribution form is graphically represented in Fig. 2.2, showing a snapshot of a student's use of GenAI for the formal analysis role in the management consulting assignment (M563). As discussed before, in student work, research activities involving generative AI should be clearly documented in line with GenAI integration rubrics and linked to the AI contribution form. Activities where AI outputs are directly used in the assignment, or partially used, or have shaped the

Table 2.9 AI contribution form for scientific work for computer science (M609) (with AI contribution assessment form instructions)

AI contribution form

Based on the assessment brief of the course "***scientific work for computer science (M609)***," you were granted permission to use the generative artificial intelligence tools like ChatGPT. However, it is essential to employ these tools to enhance learning outcomes rather than as a replacement for the manual learning efforts. In this context, these tools are meant to serve as a "***knowledge collaborator***" and should be utilized in an ethical manner.

The use of technology as a knowledge collaborator means employing it to obtain guidance, directions, or support during research or assignments without allowing it to make a substantial contribution to your final work. Ethically, the AI should not be considered a co-author of your submission *(and should not be eligible for it)*. The AI contribution assessment form specifies the different roles that technology could play in your research-driven assignment, along with guiding questions aligned with each role. This does not mean that you are required to use technology for all roles or that every broader activity suggested by individual questions must be performed. Classroom experiences with GenAI, research-based exercises, and discussions about your assignments will help you understand the possible roles, use cases, and potential misuse of the technology in the context of your research-driven assignment. The checklist can be customized—for example, by adding additional questions for different roles. Questions may also be skipped if they do not apply to the assignment or if the technology is not used for that specific role.

You are required to report your use of technology in alignment with the broader activity described by each question, along with a self-assessment of its contribution level. Please note that misreporting technology usage or its contribution levels may constitute academic misconduct, as the tutor will review the assignment, the AI contribution form, and the AI contribution assessment form to evaluate the usage and contribution of technology.

It is important to acknowledge that sometimes these tools generate wrong information. Any information obtained from such tools should be considered as merely "hypothesis." You must employ your creative problem-solving skills to improve the information generated by such tools. Your assignment will be evaluated based on its creative content and presentation, which should be driven by your creativity and problem-solving skills.

Use the AI contribution form template (***enclosed below for your reference***) to report the AI tool contribution. In particular, the AI contribution form will highlight AI tool contributions by focusing on the following aspects:

- Screenshot of the prompts given to ChatGPT.
- Which information in your research paper is provided by the AI technology (map it with the prompt information), and how much it was adapted to meet your needs or improve its accuracy?

Please keep in mind that failing to disclose the AI contribution or obscuring the actual contribution will result in an accusation of academic misconduct. We are committed to upholding academic integrity while also harnessing the full potential of technology to maximize our learning experience.

You might not need to include the specific interactions in the AI contribution form if you only utilized an AI tool as a reference and did not directly use the information in your assignment. Nevertheless, it is still a good idea to express appreciation for the tool and briefly discuss how you used it in the assignment in the acknowledgement area.

(continued)

Table 2.9 (continued)

Part A (AI-generated information was directly used and/or helped reshape the research)
The technology use involved multiple interactions, with each interaction comprising several prompts to elicit specific responses from ChatGPT, enabling a comprehensive exploration of the topic that shaped and informed the research work validated, and/or refined GenAI outputs were subsequently incorporated, either fully or partially, into the research work and the final assignment. **Student names:** [*Write here names of all students of your group*] **Project title:** [*Write here title of your project*] **Interaction 1** **Date and time:** [*Mention here the date and time of interaction 1*] **Technology role:** [This role needs to be aligned with CRediT taxonomy and GenAI integration rubrics] **Context:** [*Briefly describe why you used AI for this interaction*] **Purpose:** [*Explain the specific task or purpose for using AI*] **AI output:** [*Copy and paste or summarize the AI-generated content. Include all the outputs pertaining to the interaction*] **Student's contribution:** [*Describe how you integrated the AI-generated content into your project or analysis*] **Reflection:** [*Reflect on the value, limitations, or insights gained from this interaction. You may also focus on how the information presented by you builds on the one generated by AI tool*] **Interaction 2 (and onwards)** **Date and time:** [*Mention here the date and time of interaction 2*] **Technology role:** [This role needs to be aligned with CRediT taxonomy and GenAI integration rubrics] **Context:** [*Briefly describe why you used AI for this interaction*] **Purpose:** [*Explain the specific task or purpose for using AI*] **AI output:** [*Copy and paste or summarize the AI-generated content. Include all the outputs pertaining to the interaction*] **Student's contribution:** [*Describe how you integrated the AI-generated content into your project or analysis*] **Reflection:** [*Reflect on the value, limitations, or insights gained from this interaction. You may also focus on how the information presented by you builds on the one generated by AI tool*] **Note: Include here all the interactions that resulted in the AI-generated information as directly used in the assignment or shaped the research work.** We would like to acknowledge the use of ChatGPT as an important source of information when conducting the study for this assignment. The data produced by the application as directly used in the assignment and/or how it reshaped the research work is mentioned in Part A. This was a useful resource that helped students gain a deeper grasp of the topic.

(continued)

Table 2.9 (continued)

Part B (*AI-generated information was not included in the assignment and/or did not significantly shape the research work*)
Acknowledgement We would like to acknowledge the use of ChatGPT as a valuable informational resource during the preparation of this assignment. Although the outputs from the application were not directly included in the final work and/or had limited influence on shaping the research, they provided useful initial insights that helped the team better understand the research topic. The tool was used for a variety of things as outlined below. ***(Explain how the tool assisted you here, aligned with CRediT taxonomy and AI contribution assessment form using the following structure).*** *I acknowledge the use of ChatGPT (version XXXX) in the following roles and activities, which are also outlined in AI contribution assessment form; the outputs were not directly used in the assignment and did not reshape the research work:* ***Conceptualization:*** *[For example, assisted in generating initial ideas for research themes].* ***Formal analysis:*** *[For example, cross-checked themes generated by the team using grounded theory with GenAI-generated themes; further investigation led to the decision not to incorporate the AI suggestions].* *Add as many as roles that apply.*

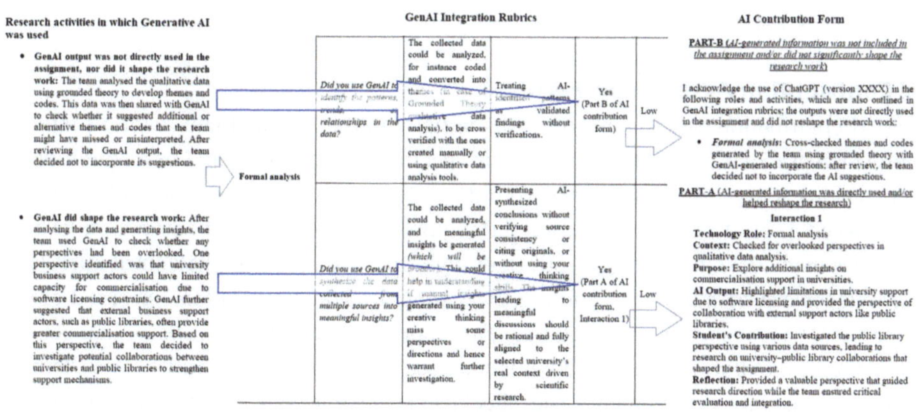

Fig. 2.2 Connection between GenAI integration rubrics and the AI contribution form for a sample example from the management consulting assignment (M563)

research work should be reported in Part A of the AI contribution form, including details of prompts, usage, and reflection on the AI's contribution as per AI course policy. Conversely, activities where AI was consulted but did not result in outputs directly incorporated into the assignment or significantly shape the research should be documented in Part B of AI contribution form, noting the role of AI in cross-checking, idea exploration, or perspective generation without direct influence on the final work.

As shown in Fig. 2.2, a student group reported the use of GenAI for formal analysis with low technology contribution levels, acknowledging the use by specifying which activity is reported in which part of the AI contribution form. For instance, in the first activity, the team analyzed qualitative data using grounded theory to develop themes and codes. GenAI was consulted to cross-check whether any additional or alternative themes were missed or misinterpreted. After reviewing the AI-generated suggestions, the team decided not to incorporate them. Since the GenAI output was not directly used in the assignment and did not shape the research work, this activity is simply acknowledged in Part B of the AI contribution form. Part B is intended for AI interactions that support or inform the research process without directly influencing the final outputs. Additionally, in the second activity, after data analysis, the team used GenAI to identify overlooked perspectives. GenAI highlighted that universities may have limitations in commercialization due to software licensing and suggested that external business support actors, such as public libraries, might provide stronger support. This insight led the team to investigate potential collaborations, directly shaping the research work and the assignment outcomes. Therefore, this activity is reported in Part A of the AI contribution form, where the students clearly documented why the technology was used, the AI output received, their own contribution, and reflections on the process. Personal details of the students and prompt screens were not included in Fig. 2.2.

AI Contribution Assessment Form

The AI contribution assessment form was initially introduced and later evolved into the GenAI integration rubric. The original form specified the role of GenAI in research-based assignments across 14 roles, aligned with the CRediT taxonomy. For each role, students were asked a set of guiding questions and required to rate GenAI's contribution level as low (less than 20%), medium (20–50%), or high (more than 50%). The contribution level percentages can be adjusted according to existing academic regulations or based on the tutor's assessment specific to the course. This approach helps students ensure that generative AI remains a knowledge collaborator and does not qualify as a co-author by systematically assessing its involvement. The checklist is flexible and can be customized, for example, by adding additional questions for different roles. Students may skip any question that does not apply to their project.

The key difference, however, is that the original form did not include role-specific examples illustrating how technology could be applied to particular questions, nor did it highlight examples where such use might cross into academic misconduct. It also did not provide guidance on whether acknowledgement of technology use was required in specific situations.

Tutors have the option of using either the AI contribution assessment form or the GenAI integration rubric. For instance, if the tutor wishes to provide students with a few assignment-aligned examples that enable controlled exploration of the technology—together with guidance on acknowledgement requirements and self-evaluation of technology use—the GenAI integration rubric is more suitable. However, if the intention is simply for students to report on the use of technology, along with the level of contribution, the AI contribution assessment form is the more appropriate choice. In either case, it is important to clearly specify the chosen approach in the AI policy within the AI contribution form. AI contribution assessment form is tabulated in Table 2.10.

Table 2.10 AI contribution assessment form *(taken from Gupta and Gupta (2024), reused with the permission of the publisher; the figure remains under the standard IEEE license)*

		Rating		
AI contribution role	Questions	Low *(less than 20%)*	Medium *(20–50%)*	High *(above 50%)*
Conceptualization	*Did you use generative AI to define the research problem or objectives?*			
	Did you use generative AI to identify the research questions or hypotheses?			
	Did you use generative AI to develop the initial idea or concept for the project?			
Data curation	*Did you use generative AI for data management tasks, for instance, organizing, cleaning, describing, enhancing, storing, and preserving data?*			
Formal analysis	*Did you use generative AI to apply statistical or computational techniques to analyze data?*			
	Did you use generative AI to identify the patterns, trends, or relationships in the data?			
	Did you use generative AI to synthesize the data collected from multiple sources into meaningful insights?			
Investigation	*Did you use generative AI to perform data collection either primary or secondary?*			
	Did you use generative AI to do a literature survey to inform the investigation?			
	Did you use generative AI to conduct experiments, surveys, or observations?			

(continued)

Table 2.10 (continued)

AI contribution role	Questions	Rating Low (*less than 20%*)	Medium (*20–50%*)	High (*above 50%*)
Methodology	*Did you use generative AI to design the overall approach or methodology for the project?*			
	Did you use generative AI to select appropriate research methods and techniques?			
	Did you use generative AI to design research study protocols and procedures for data collection and analysis?			
Software	*Did you use generative AI to develop any software application that was used in any research activity, for instance, data analysis, modeling, or visualization?*			
	Did you use generative AI to implement the algorithms or working software for specific tasks?			
	Did you use generative AI to test and debug software components that are the outcome of the research or those required by the research?			
Validation	*Did you use generative AI to check the accuracy and reliability of research findings, for instance by performing research replications?*			
	Did you use generative AI to verify the validity of assumptions and methodologies used in the project?			
Visualization	*Did you use generative AI to create or redesign the visual representations of data or research findings?*			
Writing—original draft	*Did you use generative AI to draft full or portions of the version of research papers, reports, or other project documents?*			
Writing—review and editing	*Did you use generative AI to review the research documents for clarity, coherence, and accuracy, for instance, improving expressions?*			
	Did you use generative AI to edit the document for grammar, spelling, and punctuation errors?			
	Did you use generative AI for the feedback and suggestions for improvement to enhance the quality of the writing?			

The acceptable level of the generative AI contributions in assignments should be guided by academic regulations or the tutor's own assessment of contribution percentages specific to the course. Ideally, for practical, research-oriented assignments, AI use is refined to a limited number of roles and very specific use cases, focusing on areas where it can assist without replacing core student work. For example, in the assignments outlined in Tables 2.3 and 2.4, AI use could be directional and applied to select roles such as conceptualization (e.g., generating initial ideas about challenges faced by university startup services), methodology (e.g., identifying additional data sources or exploring qualitative data analysis techniques), formal analysis (e.g., verifying themes or codes identified by the research team), investigation (e.g., clarifying difficult terms during literature review), validation (e.g., prompting AI to suggest new perspectives requiring further research), visualization (e.g., creating figures summarizing team results), and writing—review and editing (e.g., grammar corrections or improving expression). This approach ensures that AI remains a supporting tool, enhancing specific aspects of the assignment while maintaining the integrity and originality of the student's work. Table 2.11 outlines acceptable levels of GenAI

Table 2.11 Levels of collaboration and corresponding attribution for courses *(taken from Gupta and Gupta* (2024)*, reused with the permission of the publisher; the figure remains under the standard IEEE license)*

Contribution category	Major contribution roles (CRediT taxonomy)*	Outcome/Actions
Substantial	***Higher contribution of AI with respect to student contribution to either of the contributions (or a mix of these)*** *(50% or more)* Conceptualization, data curation, formal analysis, investigation, methodology, software, validation, visualization, writing (original draft), writing (review and editing)	Academic misconduct.
Moderate	***Average or less contribution of AI with respect to student contribution to either of the contributions (or a mix of these)*** *(above 20% but less than 50%)* Conceptualization, data curation, formal analysis, investigation, methodology, software, validation, visualization, writing (original draft), writing (review and editing)	Academic misconduct. Evaluate students for their contribution only (*excluding AI contribution*)
Minimal	***AI contribution as knowledge collaborator to either of the contributions (or a mix of these)*** *(less than 20% contribution)* Conceptualization, data curation, formal analysis, investigation, methodology, software, validation, visualization, writing (original draft), writing (review and editing)	AI tool acknowledgement

*The percentage contribution of GenAI across roles and possible outcomes or actions is decided by the university or the faculty

contributions and corresponding actions. Substantial contribution (50% or more) across major roles is considered academic misconduct. Moderate contribution (20–50%) is also academic misconduct, but students are evaluated based only on their own work and GenAI contributions are excluded completely. Minimal contribution (less than 20%) is acceptable, with proper AI tool acknowledgement.

Assignment Evaluation Rubrics

Assignment assessment rubrics are closely aligned with the applied research focus of the final assignments. For instance, the management course assessment rubrics were redesigned to reflect applied, real-world expectations rather than rote learning (Table 2.12). Students were evaluated on six dimensions: how well they explained the case, the quality of their strategic recommendations, the feasibility of their implementation plan, the strength of their scientific reasoning, their ability to reflect critically, and the professionalism of their formatting and referencing. Each dimension was tied to skills that AI could not replace—original analysis, creativity, feasibility judgment, and critical reflection. For example, students could use GenAI to generate ideas for possible startup services, but their grade depended on how convincingly they justified those ideas using real evidence and how feasible their implementation plan was. Practical nature of assignments with open-ended problem together with aligning grading with human judgment and creativity addressed the ability of the technology to impact academic integrity.

Table 2.12 Assessment rubrics for management consulting course (M563)

Case explanation (10%) • You will provide brief and informative information about the university selected by you as a possible case for the assignment.
Strategic recommendations (15%) • Students effectively utilize the tools and methodologies learned in the course to develop strategic recommendations for the university case. The strategic recommendations should be rational, practical, and valuable for the case.
Implementation plan (15%) • The implementation plan suggested by the students needs to be rational, practical, and feasible. It is important to consider that the team responsible for implementation *(including a consultant)* should find it easier to convert the suggested implementation plan into action.
Scientific reasoning (20%) The consulting directions and implementation plan need to be grounded in scientific reasoning. This means that students must convince the evaluator that nothing comes from random data. In fact, the technique used to disseminate information in the consulting report is based on scientific research.
Reflections (05%) • Provide your reflections that could be valuable for peer consulting teams and your case.
Professional formatting, organizing, style, and Harvard referencing (05%) • Professional report formatting, logical flow, client focus, and accurate Harvard referencing.

2.3 GenAI Integration in Courses: Summary

The discussion in this chapter demonstrates that the ethical and effective integration of generative AI in courses requires intentional redesign across teaching, assessment, and evaluation practices. By embedding AI into structured learning activities and applied research assignments, supported by clear AI policies, AI contribution form, GenAI integration rubrics, AI usage guidelines, AI contribution assessment form, and transparent reporting, institutions can safeguard academic integrity while fostering critical thinking and digital literacy through these structured safeguards. The course has been redesigned to integrate generative AI across group activities and both synchronous and asynchronous tasks, ensuring alignment with key concepts and frameworks. This supports experiential learning, allowing students to experiment responsibly, reflect on their experiences, share knowledge, and identify potential applications and misuse cases. These skills are further reinforced through challenging practical assignments and practice-focused evaluations, where technology cannot replace students' efforts. The proactive approach outlined here positions GenAI as a knowledge collaborator that strengthens student learning outcomes by cultivating knowledge, confidence, application, and awareness, without being able to make substantial contributions to the student work. These approaches help students understand the technology's limitations, recognize misuse cases, and develop strategies for responsible use of the technology. These approaches also help build the skill of reflective synthesis, that is, making informed decisions about when to engage AI and when to rely on one's own reasoning, based on critically integrating information from multiple sources, including generative AI. By evaluating and validating all inputs, including AI-generated outputs, students can use AI as an assistive tool rather than a substitute for their own efforts. The use of technology may vary across different situations, depending on students' experiences and judgments.

References

Abbas, M., Jam, F. A., & Khan, T. I. (2024). Is it harmful or helpful? Examining the causes and consequences of generative AI usage among university students. *International Journal of Educational Technology in Higher Education, 21*(1). https://doi.org/10.1186/s41239-024-00444-7

Cotton, D. R., Cotton, P. A., & Shipway, J. R. (2024). Chatting and cheating: Ensuring academic integrity in the era of ChatGPT. *Innovations in Education and Teaching International, 61*(2), 228–239. https://doi.org/10.1080/14703297.2023.2190148

Dwivedi, Y. K., Kshetri, N., Hughes, L., Slade, E. L., Jeyaraj, A., Kar, A. K., et al. (2023). Opinion paper: "So what if ChatGPT wrote it?" multidisciplinary perspectives on opportunities, challenges and implications of generative conversational AI for research, practice and policy. *International Journal of Information Management, 71*, 102642. https://doi.org/10.1016/j.ijinfomgt.2023.102642

Gruenhagen, J. H., Sinclair, P. M., Carroll, J. A., Baker, P. R., Wilson, A., & Demant, D. (2024). The rapid rise of generative AI and its implications for academic integrity: Students' perceptions and use of chatbots for assistance with assessments. *Computers and Education: Artificial Intelligence, 7*, 100273. https://doi.org/10.1016/j.caeai.2024.100273

Gupta, V. (2025). From experimentation to insight: Structured reporting of AI experimentations in libraries. *Journal of Web Librarianship, 19*(1), 93–104. https://doi.org/10.1080/19322909.2025.2450539

Gupta, V., & Gupta, C. (2024). Navigating the landscape of AI-generated text detection: Issues and solutions for upholding academic integrity. *Computer, 57*(11), 118–123. https://doi.org/10.1109/MC.2024.3445068

Gupta, V., & Nyamapfene, A. (2025). Integrating generative AI in academic assessments: Practical recommendations on AI usage and enhancing knowledge sharing. *Internet Reference Services Quarterly, 29*, 1–11. https://doi.org/10.1080/10875301.2025.2456903

Kizilcec, R. F., Huber, E., Papanastasiou, E. C., Cram, A., Makridis, C. A., Smolansky, A., et al. (2024). Perceived impact of generative AI on assessments: Comparing educator and student perspectives in Australia, Cyprus, and the United States. *Computers and Education: Artificial Intelligence, 7*, 100269. https://doi.org/10.1016/j.caeai.2024.100269

Lee, D., Arnold, M., Srivastava, A., Plastow, K., Strelan, P., Ploeckl, F., Lekkas, D., & Palmer, E. (2024). The impact of generative AI on higher education learning and teaching: A study of educators' perspectives. *Computers and Education: Artificial Intelligence, 6*, 100221. https://doi.org/10.1016/j.caeai.2024.100221

Monitoring and Evaluation (M&E) Framework

3

This chapter introduces the monitoring and evaluation (M&E) framework applied to GenAI integration, grounded in a theory of change (ToC). It begins with the problem statement—students' lack of understanding of GenAI's capabilities, limitations, and ethical dimensions—and outlines intervention strategies including practical application, collaborative learning, ethics-focused discussions, and structured evaluation. The framework defines inputs (such as training materials, AI forms, and policy briefs), outputs (student engagement, reduced misconduct cases), and outcome levels. Short-term outcomes (immediate evaluation) measure knowledge, confidence, application, and awareness directly after the course, while medium-term outcomes (follow-up evaluation) capture to measure knowledge retention, application in different contexts, long-term confidence in using GenAI, and critical awareness of technology 3–6 months later. Long-term outcomes focus on broader academic and professional adoption. The chapter also details evaluation research methodology, including sampling, data collection and analysis procedures, survey instruments, indicators, and ethical safeguards, offering peers a ready-to-use framework for evaluating GenAI interventions systematically.

3.1 Theory of Change

3.1.1 Problem Statement

Many students struggle to understand and apply GenAI tools effectively in academic work. There is often a lack of awareness regarding AI's capabilities, limitations, and ethical considerations. Without structured exposure, students may either underutilize GenAI or misuse it.

V. Gupta, C. Gupta, *Technological Innovations in Education*, Synthesis Lectures on Technology Management & Entrepreneurship,
https://doi.org/10.1007/978-3-032-11225-5_3

3.1.2 Intervention Strategy

To address these challenges, the MrciS followed a structured intervention strategy involving (a) **practical application** *(encouraging the use of GenAI in real academic work)*, (b) **collaborative learning** *(promoting group work to explore AI's potential)*, (c) **ethical considerations** *(educating students on responsible AI use)*, and (d) **structured evaluation** *(conducting surveys to measure impact)*.

3.1.3 Inputs (Resources and Activities)

- Course design integrating GenAI into group work and final exams.
- Training sessions and support materials on GenAI.
- AI contribution form for reporting the AI usage in assessments.
- AI contribution assessment sheets/GenAI integration rubrics to make the AI contribution explicit in final assessments.
- Ethics-based discussions on responsible AI use.

3.1.4 Outputs (Immediate Results)

- Students actively engage in GenAI-assisted group work and final assessments.
- Students document their contributions and AI contributions and reflect on AI use.
- Increased student participation in AI-driven learning activities.
- Reduced number of AI-related academic misconduct cases (AMCs).

3.1.5 Short-Term Outcomes (Immediate Evaluation)

These outcomes are evaluated through an immediate post-course survey and focus on:

- **Knowledge:** Improved understanding of GenAI concepts.
- **Confidence:** Increased self-efficacy in using GenAI.
- **Application:** Initial use of GenAI in coursework.
- **Awareness:** Basic awareness of AI, ethical concerns, and other technologies.

3.1.6 Medium-Term Outcomes (Follow-Up Evaluation)

These outcomes are evaluated through a follow-up survey (3–6 months later) and focus on:

- **Knowledge:** Retained and deepened knowledge about GenAI.
- **Confidence:** Sustained confidence in using GenAI beyond coursework.

- **Application:** Independent use of GenAI in academic or professional settings or application in different courses or thesis work.
- **Awareness:** Critical awareness of AI, ethics, and other relevant technologies.

3.1.7 Long-Term Outcomes (Impact)

- Widespread adoption of GenAI in academic and professional domains.
- Ethical and responsible AI use among graduates.
- Contribution to digital innovation and AI literacy.

3.2 Evaluation Details

3.2.1 Objectives of the Evaluation

The evaluation aims to assess the impact of GenAI integration in academia by systematically measuring short-term and medium-term outcomes because of intervention implemented across few degree courses and continuously improvised in next semester courses. The evaluations are conducted in two phases (which are conducted simultaneously across multiple courses—those which are just completed and those which were completed few months before). These two stages include the following:

Immediate (Short-Term) Evaluation

- Conducted immediately after the course to capture the initial effects of GenAI exposure on students.
- Focus on immediate knowledge gain, confidence, initial application, and awareness of technology *(including ethical considerations)*.

Follow-Up (Medium-Term) Evaluation

- Conducted 3–6 months after the course completion to measure knowledge retention, application in different contexts, long-term confidence in using GenAI, and critical awareness of technology *(including different related technologies)*.
- Assesses whether students continue to use and engage with GenAI beyond coursework.
- Includes questions on sustained application, ethical dilemmas faced, and whether students share their knowledge with others.

The immediate and follow-up evaluations are conducted with two objectives as explained below:

- **Immediate evaluation:** Assesses student learning, confidence, early application, and awareness of GenAI.

- **Follow-up evaluation:** Measures sustained *(or enhanced)* GenAI-related knowledge, practical usage *(or application in other contexts or areas)*, critical awareness, and confidence with the technology.

3.2.2 Evaluation Research Methodology

This evaluation employed a survey research method to assess the impact of GenAI intervention on master's degree students. The intervention was implemented in selected courses where GenAI tools were integrated to enhance learning. The impact evaluation, covering both short-term and medium-term impacts, employed surveys to collect data from students enrolled in two courses: management consulting (M563) was taught during the summer semester 2024 *(April to July 2024)* to evaluate medium-term impact, while project communication and stakeholder management (M562) was taught in the winter semester 2025 *(January to March 2025)* to assess short-term impact. The short-term impact evaluation of M562 involved feedback gathering immediately after the course. The medium-term impact evaluation of M563 involved collection of data as part of the evaluation several months after course completion. Although the immediate and follow-up evaluations involved different student groups, both cohorts participated in courses designed with a comparable level of assessment complexity and a consistent approach to GenAI integration. In each course, students received the same guidance on how to use GenAI during and beyond classroom activities, and they were required to submit the same documentation if they chose to use GenAI in their final assignments. As a result, the evaluation outcome can be viewed as complementary perspectives, offering insight into how students at different points engaged with GenAI under a shared instructional and assessment framework.

The evaluation involved the survey that aims to capture both short-term and medium-term outcomes related to students' knowledge, confidence, application, and awareness of GenAI, including ethical considerations. The four aspects—knowledge, confidence, application, and awareness—are utilized in the ALA Project Outcome Toolkit (https://www.projectoutcome.org/) for conducting M&E in libraries to assess the impact of their programs, including education and learning. These aspects were also employed in real-world settings, where ChatGPT Business programs were evaluated for both short-term and long-term impact by Gupta (2025a, 2025b). The same evaluation dimensions have been applied in assessing the impact of courses M562 and M563, the details of which, including the results, are disseminated in this book.

Participants

The participants are master's degree students enrolled in courses M562 and M563, representing various degree programs and nationalities. They were engaged with generative AI tools (ChatGPT) as part of their coursework at Gisma University, with both courses taught by the Director of MrciS. The study did not collect any personal identifiable information to maintain privacy and confidentiality, ensuring that responses are anonymous.

Data Collection

Data was collected using a questionnaire designed to assess the immediate and medium-term impacts of GenAI intervention *(Tables 3.1 and 3.2)*. The survey link was administered to participants of the two courses from 1 April to 15 May 2025. The survey included both Likert-scale questions and open-ended questions. The survey questions assessed students' responses to various statements about their knowledge, confidence, application, and awareness of GenAI, using the 5-point Likert scale (ranging from "1: Strongly Disagree" to "5: Strongly Agree"). The survey questions were aligned with the indicators that measure the outcomes in the following four aspects: *knowledge, confidence, application, and awareness*. The survey questionnaires differed for the immediate and follow-up evaluations, as the formulation of these four aspects will change to reflect the different objectives and time frames of the evaluations. A few optional open-ended questions were included to provide qualitative insights to complement the quantitative data and will be tailored to address both immediate and long-term impacts, allowing for more detailed reflections.

Data Analysis

The data collected through the Likert-scale responses was analyzed using descriptive statistics, such as mean, and percentages. This helped in understanding the overall trends in students' knowledge, confidence, application, and awareness of GenAI both immediately after the intervention and in the medium term. For the open-ended responses, qualitative analysis was performed using grounded theory (GT) to identify recurring themes and insights related to students' perceptions across the four aspects. This analysis provided richer, context-specific information to complement the quantitative data.

Table 3.1 Outcome aspect, indicators, survey questions for immediate evaluations

Aspect	Indicator	Survey questions
Knowledge	% of students correctly identifying GenAI usage, strengths, and limitations	How well do you understand GenAI capabilities and limitations after completing the course? (Likert Scale)
Confidence	% of students confident in using GenAI for coursework	How confident are you in using GenAI for academic work? *(Likert Scale)*
	% of students comfortable explaining GenAI to peers	How confident are you in explaining GenAI to the peers? *(Likert Scale)*
Application	% of students who used GenAI in final assessments *(individual or in group)*	Did you use GenAI in your final assessment? *(Binary—Yes or No)*
Awareness	% of students recognizing basics of ethical considerations	How much basic awareness of ethical AI use you gain during your course? *(Likert Scale)*
	% of students recognizing at basic level the existence of other GenAI technologies	Did you become aware of the existence of other GenAI technologies? *(Binary—Yes or No)*

Table 3.2 Outcome aspect, indicators, survey questions for immediate evaluations

Aspect	Indicator	Survey questions
Knowledge	% of students who recall and apply key GenAI concepts *(beyond what was discussed in the classroom)*	How much new have you learned about GenAI-related concepts you have learned through selfstudy, research, or practical application since the course? *(Likert scale)*
Confidence	% of students confident in using GenAI for professional or research tasks	How much confidence have you gained in using GenAI for the tasks that required complex problem-solving—where solutions couldn't be generated solely by technology, but also needed research strategies, critical thinking, and external data? *(Likert scale)*
	% of students who teach/share GenAI knowledge with others	How much GenAI learning support have you provided to your colleagues, classmates, or friends? *(Likert scale)*
Application	% of students using GenAI for research, job tasks, creative projects, or other courses?	Do you still use GenAI? *(Binary—Yes/No)*
		If yes, in what ways have you applied it (e.g., research, work, other courses, personal projects)? *(Open-ended)*
Awareness	% of students advocating responsible AI practices	Have you encouraged others to use GenAI responsibly? *(Binary—Yes/No)*
	% of students using other GenAI technologies	Have you used other GenAI technologies, apart from ChatGPT? *(Binary—Yes/No)*

Ethical Considerations

The survey ensured that students' personal information was not collected, maintaining anonymity and confidentiality. Participation was voluntary, and students were informed that they could skip any question, including the open-ended ones, without penalty. The study adhered to ethical guidelines to protect participants' rights and ensure that data be handled responsibly.

3.2.3 Outcome Aspect, Indicators, and Survey Questions

For Immediate Evaluations

The indicators to measure the outcomes across the four aspects and the associated survey questions for immediate evaluations are tabulated in Table 3.1.

For Follow-Up Evaluations

The indicators to measure the outcomes across the four aspects and the associated survey questions for follow-up evaluations are tabulated in Table 3.2.

Tables 3.1 and 3.2 present only the main questions from the immediate and follow-up survey questionnaires. The questionnaires for both the short-term and medium-term impact evaluations are provided in Appendix III and Appendix IV. These questionnaires are the revised versions developed after pretesting and were used to evaluate the impact of GenAI integration in courses on students' knowledge, confidence, application, and awareness in the short and medium term. Details of the pretesting process are discussed in Chapter 4, while the questionnaires for short-term and medium-term impact evaluations subjected to pretesting are included in Appendices I and II.

References

Gupta, V. (2025a). Factors influencing librarian adoptions of ChatGPT technology for entrepreneurial support. *Library Hi Tech.* https://doi.org/10.1108/LHT-09-2024-0463

Gupta, V. (2025b). From experimentation to insight: Structured reporting of AI experimentations in libraries. *Journal of Web Librarianship, 19*(1), 93–104. https://doi.org/10.1080/19322909.2025.2450539

Pretesting of Survey Instruments: Methodology and Results 4

This chapter outlines the methodology and results of the pretesting phase conducted to refine the survey instruments used for evaluating the integration of generative AI (GenAI) in the master's courses M562 and M563. Pretesting was a crucial step to ensure that the immediate and follow-up evaluation questionnaires met high standards across several dimensions: clarity and comprehension of questions, appropriateness and acceptability of response options, impartiality and lack of bias, overall quality of the questionnaire, clarity and sufficiency of instructions, logical flow of questions, and time required to complete the survey. By involving a diverse group of master's students and using both quantitative feedback and follow-up interviews, the process identified ambiguities, cultural nuances, and practical improvements to strengthen the instruments. The refinements and insights presented in this chapter aim to support peers who wish to reuse or adapt the pretesting approach and survey instruments for similar educational interventions involving GenAI or related technologies.

4.1 Pretesting of Survey Research Instrument

Before administering the final surveys for the immediate and follow-up evaluations, a pretesting phase was conducted to identify and address potential issues with the instrument. This phase ensured that problems related to the logical flow, clarity, comprehension, and impartiality of survey questions and others could be detected early. Any issues identified during pretesting were corrected before the full-scale survey administration. Additionally, pretesting helped assess whether the chosen data collection method (i.e., surveys in the current case) was suitable for evaluating both the immediate and follow-up impacts.

V. Gupta, C. Gupta, *Technological Innovations in Education*, Synthesis Lectures on Technology Management & Entrepreneurship,
https://doi.org/10.1007/978-3-032-11225-5_4

4.2 Pretesting Research Methodology

The pretesting phase involved five master's students who had completed a course on the ethical integration of GenAI (though this course itself was not part of the evaluation). These students were selected to ensure diversity in cultural backgrounds, language proficiency, and academic disciplines. Additionally, they had varying academic performances in the course, ranging from the 60s to above 90. The participants for the pretesting phase were identified based on the details available in the university's Canvas system for the course they had studied. A purposive sampling approach was used, ensuring that the selected students represented diversity.

Data for the pretesting phase was collected by administering both the immediate and follow-up survey questionnaires separately, as presented in Tables 4.1 and 4.2 (survey as data collection method) in the second week of March 2025. In addition, new questions were included to gather feedback on the survey's instruments. The questionnaire used for pretesting is provided in Appendix I. The data collected from the pretesting was analyzed using descriptive statistics for the quantitative (yes/no and Likert scale) responses. For the qualitative open-ended responses, a grounded theory approach was applied to identify recurring themes and insights.

To gain deeper insights into participants' responses and any additional information they wished to share, a follow-up interview was conducted. Individual information was not recorded, and informed consent was obtained before participants began filling out the pretesting questionnaire (Appendix I for pretesting of immediate survey research instrument and Appendix II or follow-up survey instrument). Only aggregated responses were used in the analysis, with selected quotes included in the report after removing any personally identifiable information. Individual responses were retained for a period of 1 month following the completion of the pretesting phase, after which they were securely deleted.

The pretesting focused on several key factors, including the *clarity and comprehension of survey questions from the participants' perspective; the appropriateness, impartiality, and acceptability of response options; as well as the overall quality of the questionnaire.* It also assessed the *clarity and sufficiency of the instructions provided, the logical flow of survey questions, and the time required* to complete the survey. These aspects were outlined by Gupta (2025a, 2025b). These additional questions were integrated into each evaluation aspect of the questionnaire, specifically addressing knowledge, confidence, application, and awareness. These questions are added to the survey to conduct the pretest (Table 4.1).

Table 4.1 Pretesting survey's additional questions for each evaluation aspect for both immediate and follow-up survey

Pretesting criteria *(for all questions against each evaluation aspect)*	Additional survey questions *(for both immediate and follow-up evaluations)*
Clarity and comprehension	*Did you clearly understand all the questions in this survey? (Yes/No)* *If No, please specify which questions were unclear and why. (Open Question)*
Appropriateness of response options	*Did the response options provided for each question allow you to express your views accurately? (Yes/No)* *If No, please suggest any improvements. (Open Question)*
Impartiality and lack of bias	*Did you find any survey questions or response options leading, biased, or suggesting a particular answer? (Yes/No)* *If Yes, please explain. (Open Question)*
Acceptability of response options	*Were there any response options that you found confusing, inappropriate, or difficult to choose from? (Yes/No)* *If Yes, please provide details. (Open Question)*
Overall quality of the questionnaire	*Would you consider this questionnaire to be well structured, readable, and relevant? (Yes/No)* *If No, please provide any suggestions for improvement. (Open Question)*
Clarity and sufficiency of instructions	*Were the instructions provided in the questionnaire clear and sufficient to guide you through the survey? (Yes/No)* *If No, what additional guidance would have been helpful? (Open Question)*
Logical flow of questions	*Did the sequence of questions feel logical and easy to follow? (Yes/No)* *If No, please specify any abrupt transitions or inconsistencies. (Open Question)*
Time required to complete the survey	*Did you feel that the time required to complete the survey was appropriate? (Yes/No)* *If No, please specify whether it was too long or too short and suggest an ideal duration. (Open Question)*

4.3 Results of Pretesting: Adjustments Made to Survey Instruments

The pretesting phase provided valuable insights into how students from diverse cultural, academic, and linguistic backgrounds interpreted the survey questions. The results revealed several key challenges, ranging from question clarity and comprehension to differences in ethical AI interpretations across students. By analyzing both quantitative feedback (Likert-scale and yes/no responses) and qualitative insights (open-ended reflections), it became evident that some questions needed refinement to improve their accuracy, inclusivity, and effectiveness. The pretesting phase identified critical areas for refinement, ensuring that the final survey would provide a more accurate and representative assessment of students' engagement with GenAI. The following section details the specific pretesting survey participants' response trends, key issues reported, and changes made to survey instruments. The final questionnaire for undertaking the immediate and follow-up survey is provided in Appendixes III and IV.

4.3.1 Common Improvements to Immediate (Short-Term) and Follow-Up (Medium-Term) Evaluation Survey

The three changes suggested for both survey instruments include (a) adding brief introduction to four aspects in the survey, (b) adding examples in each section of the questionnaire representing one of the four evaluation aspects, and (c) adding more open questions for deeper insights.

Adding Brief Introduction to Four Aspects in the Survey

Before participants begin answering the questions, a short introduction should be provided, explaining the four aspects (knowledge, confidence, application, awareness) in simple terms. A new introductory paragraph is added for each aspect in each survey instrument for the immediate evaluation survey and follow-up survey.

Introductory Paragraphs in Immediate Survey

Knowledge Aspect

This survey measures the impact of the GenAI course on your learning, confidence, and awareness of AI tools immediately after getting exposed to the technology in the course. The knowledge aspect measures your understanding of GenAI concepts, for instance how to write prompts; its capabilities, for instance what it can do for you; and its limitations, for instance wrong outputs. The idea here is to see how much basic understanding you have that helps you to understand from a user perspective about what is generative AI, how does it work, how to get the output, and what could be the issues with the technology, at basic interaction levels. It assesses your ability to recognize key terms and fundamental principles.

Confidence Aspect

This survey measures the impact of the GenAI course on your learning, confidence, and awareness of AI tools immediately after getting exposed to the technology in the course. The confidence aspect measures how comfortable and self-assured you feel when using or explaining GenAI to your peers, at basic levels. This means that you may not be an expert but feel some confidence to think about using it or to discuss about this technology. At basic levels, it *measures increased self-efficacy in using GenAI for simple tasks and assesses your comfort in trying out GenAI tools.*

Application Aspect

This survey measures the impact of the GenAI course on your learning, confidence, and awareness of AI tools immediately after getting exposed to the technology in the course. The application aspect measures students' initial use of GenAI in coursework and assesses whether they have attempted to apply GenAI in assignments or to explore the course topics without using it for final assessments. It evaluates your ability to integrate basic GenAI outputs into the academic work—seeking further directions, clarity of topics, etc.

Awareness Aspect

This survey measures the impact of the GenAI course on your learning, confidence, and awareness of AI tools immediately after getting exposed to the technology in the course. The awareness aspect measures your basic understanding of ethical AI use and availability of other GenAI technologies.

Introductory Paragraphs in Follow-Up Survey

Knowledge Aspect

This survey measures the impact of the GenAI course on your learning, confidence, and awareness of AI tools after few months of getting exposed to the technology in the course. The knowledge aspect measures retained and deepened knowledge about GenAI. It assesses the ability to recall key concepts and in being in processing of more knowledge about the technology beyond what was acquired in the course. It also evaluates progress in grasping GenAI's evolving capabilities and limitations.

Confidence Aspect

This survey measures the impact of the GenAI course on your learning, confidence, and awareness of AI tools after few months of getting exposed to the technology in the course. The confidence aspect measures your enhanced confidence levels in using GenAI beyond coursework, possibly for more complex tasks, and describing this technology to your peers, for instance making them aware and helping them get better understanding of the technology through practical use. This measure is not about application, i.e., did you really apply it, but about how confident you feel in doing so.

Application Aspect

This survey measures the impact of the GenAI course on your learning, confidence, and awareness of AI tools after few months of getting exposed to the technology in the course. The application aspect measures independent use of GenAI in academic or professional settings. It assesses application in different courses, thesis work, or job-related tasks. It also evaluates the ability to leverage GenAI for complex problem-solving and decision-making.

Awareness Aspect

This survey measures the impact of the GenAI course on your learning, confidence, and awareness of AI tools after few months of getting exposed to the technology in the course. The awareness aspect measures critical awareness of AI, ethics, and other relevant technologies. This means that you should now be fully aware of various ethical aspects, and maybe you have experimented with other GenAI technologies.

Adding Examples in Each Section of the Questionnaire Representing One of the Four Evaluation Aspects

Several students in the pretesting phase suggested adding examples to each section of the questionnaire. However, since students are exposed to these aspects throughout their course, examples are unnecessary to make the sections more informative. Additionally, including examples could introduce bias, as respondents may focus only on the given examples rather than considering their unique experiences. A more neutral wording ensures that responses reflect actual behavior rather than being influenced by survey phrasing. Furthermore, the introductions to each questionnaire section pertaining to each evaluation aspect have been redrafted to clearly communicate what each section is measuring.

Adding More Open Questions for Deeper Insights

The pretesting revealed that some students felt the survey was too rigid and wanted opportunities to explain their perspectives. To address this, one open-ended question was added to each section. Table 4.2 tabulates the new open questions that were added to immediate and follow-up surveys.

Table 4.2 Open-ended questions added to the questionnaire

Evaluation aspect	Immediate survey questionnaire	Follow-up survey questionnaire
Knowledge	What is one key GenAI concept you learned in this course that you did not know before?	What new concepts about GenAI have you learned beyond what was explored during your course?
Confidence	Could you please elaborate on your confidence in using or discussing this technology with your peers, at basic levels?	Could you please elaborate on your confidence in using or discussing GenAI with your peers, particularly at a level beyond basic understanding? .
Application	Can you describe a specific task where GenAI was most helpful to you during the course, regardless of whether you used it for the final assessment?	Can you describe a specific task (e.g., tasks pertaining to *(but not limited to)* research, work, other courses, personal projects) where GenAI was particularly helpful to you in the time post-course completion?
Awareness	Can you share any ethical concerns while using GenAI in your course? Can you share any other GenAI technologies you became aware of during the course?	Can you share an instance where you or someone you know faced an ethical dilemma related to GenAI? Which other GenAI technologies did you try experiment with post-course?

4.3.2 Improvements to Immediate (Short-Term) Evaluation Survey

- **Enhancing Knowledge Section for Clarity**
  The **knowledge-based questions** were found to be **too broad** by the students who expressed their disagreements as response to question clarity and comprehension, appropriateness of response options, and clarity and sufficiency of instruction questions (questionnaire aspect-related section). Students from nontechnical disciplines expressed uncertainty about whether they should rate themselves based on theoretical understanding or their ability to use GenAI in practice. A key improvement was made by modifying the question ***"How well do you understand GenAI after completing the course?"*** to ***"How well do you understand GenAI's capabilities and limitations after completing the course?"*** This change makes the question more precise and avoids confusion regarding technical depth vs. practical application. One student remarked in the follow-up interview, "*I understand how to use GenAI, but I don't know how it works behind the scenes. Should I rate myself lower?*"
- **Refining Confidence Section to Capture More Nuanced Responses**
  The confidence-related questions showed inconsistent self-rating patterns, with most of the students hesitating on how to evaluate themselves. Some students, especially those with lower academic performance (60s–70s range), reported feeling "not confident enough" even if they used GenAI regularly. The follow-up open-ended question was added in the confidence section of the questionnaire, "*Could you please elaborate on your confidence in using or discussing GenAI with your peers, particularly at a level beyond basic understanding?*" (please refer to Table 4.2 for details). This additional question helps identify barriers to confidence, such as lack of technical understanding, fear of reliance on AI, or concerns about ethical issues.
  One student stated as response to the open question of the pretesting survey, "*I used GenAI in my assignments, but I still feel unsure about its correctness.*" A similar issue was found in the question about explaining GenAI to others, with disagreed students found "explaining" unclear—some interpreted it as casual discussion, while others thought it referred to teaching in-depth AI concepts. The question was reformulated to ***"How comfortable are you in describing GenAI's use to classmates?"*** This revision avoids the implication of formal instruction while still capturing students' confidence in AI discussions.
- **Broadening Application Section to Capture More Scenarios**
  One of the most frequent concerns as reported through response to the open question and in the follow-up was that the question ***"Did you use GenAI in your final assessment?"*** (Yes/No) was **too restrictive**. Several students used GenAI for coursework but **not** in their final exam, which led them to answer "No" even though they had engaged with the technology. The question was revised to ***"Did you use GenAI in any coursework or final assessment?"*** This modification ensures that students who applied GenAI in any aspect of their academic work can accurately report their usage.

4.3.3 Improvements to Follow-Up (Medium-Term) Evaluation Survey

- **Refining Knowledge Section for Clarity on "New Learnings"**
  The question "*How much new have you learned about GenAI since the course?*" confused students, who were not sure if "new" knowledge had to come from self-study, external resources, or workplace experience. This question was redrafted to the better version as "*Since completing the course, how much new GenAI-related concepts you have learned through self-study, research, or practical application?*" This revision clearly defines where new knowledge could come from, reducing ambiguity. A student commented, "*I didn't take another AI course, but I read some articles. Does that count?*"
- **Modifying Confidence Section to Specify "Complex Tasks"**
  35% of students were not sure what counted as a "complex task" in the question "*How much confidence have you gained in using GenAI for complex tasks related to work or research?*" The question was redrafted as "How confident are you in using GenAI for the tasks that required complex problem-solving—where solutions couldn't be generated solely by technology, but also needed research strategies, critical thinking, and external data?" This ensures that students understand the intended meaning of "complex tasks" without confusion.
- **Expanding Application Section to Allow for More Detailed Responses**
  The binary "yes/no" question "Do you still use GenAI?" was considered too restrictive by most of the students. Many wanted to explain how they use it. An open-ended question was added *(please refer to Table 4.2)*. The need for adding open-ended questions for all the sections of the questionnaire was mentioned by many students. A student noted, "*I still use AI, but only for brainstorming. That's different from using it for coding.*"

4.3.4 Results of Pretesting for Immediate and Follow-Up Questionnaire (Questionnaire Aspect Section Only)

The aggregated responses for the questionnaire aspect section for immediate survey instruments are graphically presented in Fig. 4.1, and follow-up survey instruments' pretesting results in Fig. 4.2. The pretesting responses for the immediate and follow-up survey research instrument appear favorable, except for certain concerns related to question clarity and comprehension, appropriateness of response options, and clarity and sufficiency of instructions. About 80% of students agreed that the time required to complete the survey was sufficient in pretesting both research instruments. The disagreement from one student may have been due to difficulties in understanding certain questions, which extended their completion time. For pretesting, agreement is represented by Likert scale values 4 and 5, and disagreement by Likert scale values 1, 2, and 3. The decision to include a value of 3 in disagreement was intentional, allowing room for potential improvement to the maximum levels.

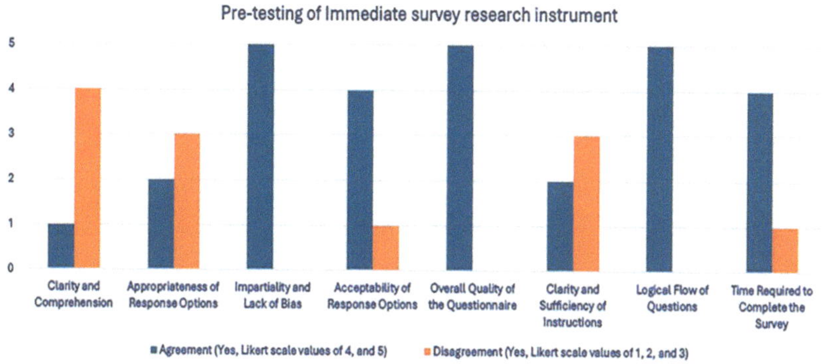

Fig. 4.1 Aggregated responses for the Questionnaire aspects (immediate evaluation survey instrument)

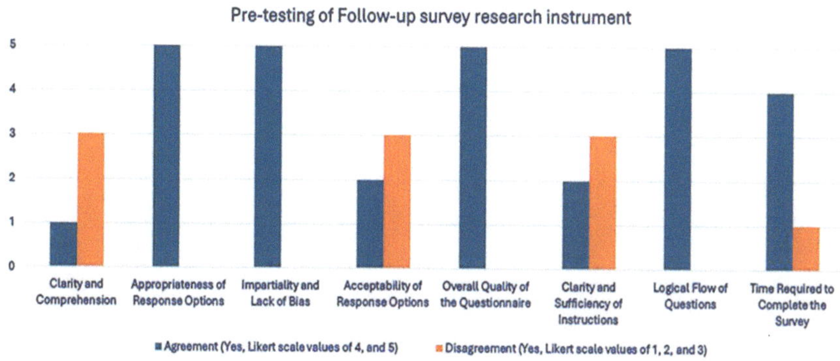

Fig. 4.2 Aggregated responses for the Questionnaire aspects (follow-up evaluation survey instrument)

Based on participant feedback from open-ended responses and follow-up interviews, key issues included the need for clearer terminology, redrafting ambiguous questions, and providing more response options (particularly open-ended questions). These concerns likely contributed to the reported disagreements. The feedback from the students has been incorporated, and research instruments are modified accordingly (Appendixes III and IV).

References

Gupta, V. (2025a). Factors influencing librarian adoptions of ChatGPT technology for entrepreneurial support. *Library Hi Tech*. https://doi.org/10.1108/LHT-09-2024-0463

Gupta, V. (2025b). From experimentation to insight: Structured reporting of AI experimentations in libraries. *Journal of Web Librarianship, 19*(1), 93–104. https://doi.org/10.1080/19322909.2025.2450539

5 Evaluation Results: Analysis of Outcomes and Impacts

This chapter presents the results and analysis of the evaluations conducted to assess the integration of generative AI (GenAI) in the master's courses M562 and M563. Building on the monitoring and evaluation (M&E) framework, the evaluations measured the short-term and medium-term impacts of the GenAI intervention across four outcome aspects: knowledge, confidence, application, and awareness. The immediate evaluation focused on students in M562 at the end of the course, while the follow-up evaluation drew upon responses from a different cohort enrolled in M563 several months after course completion. This two-phase design provides complementary insights into how students' knowledge, confidence, application, and awareness appeared at different stages of similar interventions. By combining quantitative results with qualitative reflections, this chapter offers evidence to support peers in reusing or adapting the evaluation approach for comparable educational contexts.

5.1 Immediate Evaluation Results: Analysis of Outcomes and Impacts

This section presents the immediate (short-term) evaluation results, conducted directly after the completion of selected academic courses where the GenAI intervention was implemented. The purpose of this evaluation is to assess the immediate impact of integrating generative AI (GenAI) into course delivery, focusing on the short-term outcomes as defined in the monitoring and evaluation (M&E) framework.

The immediate evaluation is designed to capture the initial effects of students' exposure to GenAI during their courses that integrated the GenAI. This includes evaluating changes in:

V. Gupta, C. Gupta, *Technological Innovations in Education*, Synthesis Lectures on Technology Management & Entrepreneurship,
https://doi.org/10.1007/978-3-032-11225-5_5

- **Knowledge**: students' understanding of GenAI's core concepts, capabilities, and limitations
- **Confidence**: students' perceived self-efficacy and comfort level in using GenAI and discussing it with peers
- **Application**: students' early use of GenAI in coursework, either in assignments or in assessments
- **Awareness**: students' initial recognition of ethical concerns and understanding of the broader landscape of GenAI tools

As described in the M&E framework, the objective of the immediate evaluation is to assess student learning, their confidence in engaging with GenAI, the initial application of GenAI in academic settings, and their awareness of associated technologies and ethical considerations. This evaluation provides insights into how effectively the course interventions initiated the learning process and influenced behavior at the end of the intervention.

Data for this evaluation was collected through a structured questionnaire designed around these four aspects, incorporating Likert scale questions, binary responses, and open-ended reflections. The analysis that follows presents both quantitative and qualitative insights, enabling a rich interpretation of student experiences and the early outcomes of the intervention. The following sections present detailed findings across the four M&E aspects.

5.1.1 Knowledge: Understanding GenAI Concepts

This aspect assesses the extent to which students developed a foundational understanding of GenAI after completing the course, particularly its capabilities, limitations, and role in academic tasks. According to the survey:

- 10% of students selected 5 (Strongly Agree).
- 55% selected 4 (Agree).
- 30% selected 3 (Moderate Agreement).
- Only 5% selected 2 (Disagree).

Figure 5.1 shows that 95% of respondents rated themselves at level 3 or above, indicating that nearly all students developed at least a moderate understanding of GenAI concepts, with a strong majority (65%) indicating clear agreement or strong agreement (Likert value 4 and 5). This suggests that the course successfully introduced core principles, especially regarding the functional understanding of GenAI tools, their capabilities, and limitations. The 5% who selected value 2 may reflect students who had difficulty grasping the abstract or technical dimensions of GenAI, perhaps due to limited prior exposure.

Open-ended reflections reinforced these quantitative trends. Notable insights about GenAI concepts learned by the students during the course include the following:

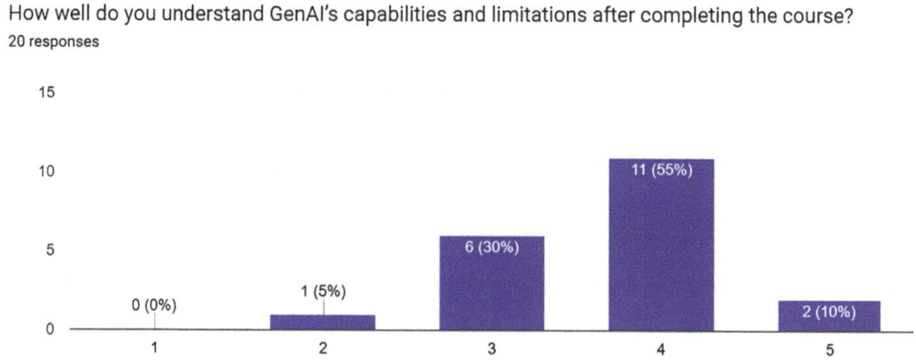

Fig. 5.1 Evaluation results for immediate survey for knowledge aspect

- *I learned about "hallucination"—that GenAI can generate completely false information that sounds plausible. I did not expect that it could just make up facts, so now I always check outputs. There is no substitute to the traditional research methods.*
- *Empathy and the human touch in certain aspects became much clearer after this course. I previously assumed that providing detailed specifics would guarantee the expected output, but I learned that this is not necessarily the case.*
- *Its functionality, limits of usage, and ethical issues. The course do covered almost everything, a nice, structured framework to explore technology within the safeguards.*
- *What could be possible use cases and why not to rely on it completely.*
- *One key GenAI concept I learned during my project communication and stakeholder management course was the importance of stakeholder-specific messaging. The course's GenAI guidelines demonstrated how to experiment with the technology effectively. I found that combining information from multiple sources and prompting the technology more narrowly help generate more precise directions. This enabled me to formulate meaningful, clear communication tailored to each stakeholder's priorities, concerns, and technical background. In other words, it highlighted that effective prompting requires both subject matter understanding and insights drawn from diverse data sources to obtain useful guidance from the technology.*
- *GenAI not only generates responses to the prompts it receives but can also create content on its own, which often lacks validation. I also learned that prompts need to be clear and precise to obtain better, more reliable answers.*
- *Functionality, limitations, prompting, experimentation, and role of data literary skills.*

The analysis of qualitative responses suggests that students initially gained fundamental knowledge of GenAI's capabilities and limitations, particularly concepts such as hallucinations, biases, and need for validation. For instance, some noted that they became aware that AI could "just make up facts," which prompted them to cross-check outputs with traditional research methods. This outcome links directly to the GenAI integration approaches presented in Chap. 2, especially embedding GenAI into group tasks where outputs were marked as correct or incorrect and validated through peer discussion, technology usage reflections, and detailed GenAI usage instructions provided. The need for continuous experimentation with the technology and need for data literary skills were reported in the survey response. The overall findings for the "knowledge" aspects align with the M&E goal of enabling baseline conceptual clarity immediately post-intervention.

5.1.2 Confidence: Self-Efficacy in Using and Explaining GenAI

The confidence dimension evaluates students' perceived self-efficacy in engaging with GenAI and their ability to explain its use to peers. In particular, two questions evaluated students' confidence: one on using GenAI for academic tasks and another on describing its use to classmates. According to the survey:

Confidence in Using GenAI (Q1)

- 15% selected 5 (Strongly Agree).
- 30% selected 4.
- 45% selected 3.
- 10% selected 2.

Confidence in Explaining GenAI to Peers (Q2)

- 15% selected 5 (Strongly Agree).
- 35% selected 4.
- 40% selected 3.
- 10% selected 2.

Figures 5.2 and 5.3 show that across both questions, 90% of students rated themselves at 3 or above, indicating a generally positive sense of self-efficacy. The distribution suggests that students feel confident engaging with GenAI personally, as well as explaining it to peers. The reasons could be that experimentations with technologies like ChatGPT are fostered through collaboration and knowledge exchanges with peers.

Open-ended reflections about the student's confidence with the technology reinforced these quantitative trends. Notable insights include the following:

- *I am becoming more confident, especially after working in a group where each of us experimented with ChatGPT for different tasks in the group assignment. I can now explain its basic functions and risks, but I still hesitate to use it for formal parts of assignments, such as certain aspects of research.*
- *Confidence comes from our understanding we got in course and what we discussed as part of our group work and as part of our final assignments. This is definitely worth sharing with those looking for the discussion.*

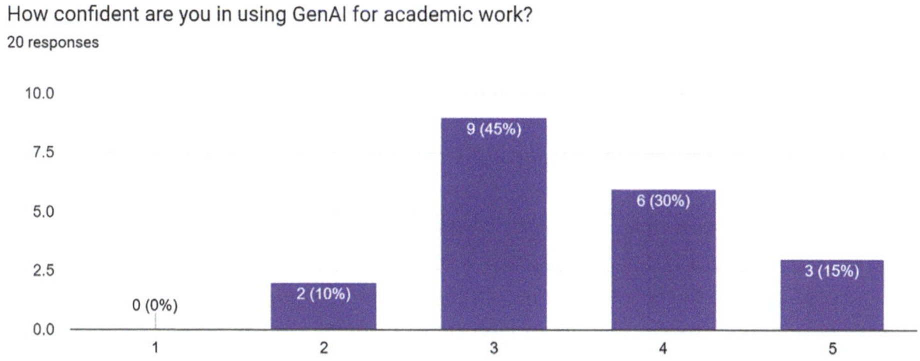

Fig. 5.2 Evaluation results for immediate survey for confidence aspect (Q1)

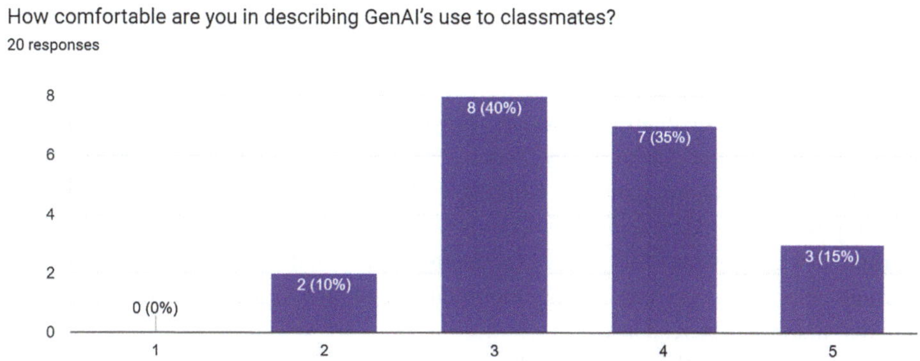

Fig. 5.3 Evaluation results for immediate survey for confidence aspect (Q2)

- *There are certain elements in GenAI-generated outputs that are easy to recognize if one has used the technology. I am becoming confident in using GenAI to its full potential and discussing it with my peers. While GenAI can provide a solid theoretical foundation on widely available topics, its practical application often requires fine-tuning with the knowledge and expertise of individuals or groups in the respective areas.*
- *I can explain the basics that we learned in the course and also what we learned in our group work, especially following the GenAI usage guidelines. That was useful to boost our confidence to experiment more with this technology.*
- *I can explain the technology and the way anyone should experiment with it to find some directions for finding solutions to the complex problems.*
- *It can be used to increase work efficiency and reduce effort, for instance automating routine tasks. Moreover, I hope and believe that it can be applied for marketing, information gathering, and many other emerging uses. I am confident in sharing my understanding of these applications with anyone as I used this technology to create social media messages.*

- *I use this technology as guidance that provides me with clues to navigate my research process, and I am gaining more confidence to explore it further. The GenAI integration rubrics are very useful here. I am confident to share my experiences about the research activities where this technology could be practically applied.*
- *My peers attended all sessions and were exposed to this technology as part of the course tasks. It was good to enhance our confidence as groups are very diverse, with few members having good experience in industries. We were able to discuss about limits of usage of the technology and areas where potentially it will not work.*
- *I am confident that I can explain the technology using simple terms, analogies, or real-world examples, especially to peers who might not have a deep technical background.*
- *My group encourages questions and learning. This makes me more comfortable sharing what I know and asking about what I do not.*
- *I feel at ease talking about basic aspects, such as crafting effective prompts and using AI responsibly. I also enjoy showing my friends how to apply it for simple tasks like writing, creating images, or brainstorming ideas.*

The analysis of qualitative responses suggests a trajectory of growing confidence. Immediately after the course, many students pointed to group activities and structured rubrics as supportive mechanisms: working collaboratively appeared to reduce anxiety and promote active experimentation. Comments such as becoming "more confident after working in a group where each of us experimented with ChatGPT" reflect the role of peer learning emphasized in Chap. 2. The important role of GenAI usage guidelines was also reported in the survey response.

This aligns well with the framework's intention to measure early-stage confidence—a foundation on which deeper comfort and articulation skills can be built over time. The presence of 10% at level 2 signals that a small group may require additional support or peer-led guidance.

5.1.3 Application: Initial Use in Academic Work

This aspect captures whether students actively applied GenAI tools during their coursework or assessments. When asked whether they had applied GenAI in their coursework or assessments, the survey results indicate the following:

- 55% answered yes.
- 45% answered no.

At first glance, this split may suggest only a modest level of application; however, a deeper understanding of the course structure reveals a more nuanced picture. The course explicitly allowed the use of ChatGPT as a "knowledge collaborator" in final assignments, and in-class usage and group work were central components of the intervention. Students who responded “no” typically referred to not using GenAI in their home tasks on the online learning system or final graded assignments. However, most of these students actively engaged with GenAI tools during classroom sessions, especially within group-based collaborative work, brainstorming tasks, and peer learning activities.

Figure 5.4 shows that more than half of the cohort (55%) formally applied GenAI in online learning system tasks or final assessments, but the real extent of engagement is underrepresented in binary yes/no reporting. The classroom use may have been treated as informal but was widely adopted as revealed at the time of group presentations after each session. The use of GenAI during group activities and other tasks in classroom sessions may not have been reported as formal application, as these activities were not directly linked to graded coursework or formal assessment submissions. It is also important to note that the final assignment was complex, requiring creativity, critical problem-solving skills, and independent research. As such, the decision to integrate GenAI into the final product often depended on the group’s shared judgment about whether the technology could meaningfully contribute to such high-level academic tasks.

Another possible explanation for the 45% of students who reported not using GenAI in formal coursework may be a lingering fear of academic misconduct. Despite the course framing GenAI as a knowledge collaborator, some students may have chosen to avoid using it in graded tasks due to concerns about ethical boundaries or to avoid any risks with the technology usage owing to technology limitations they got aware of during experimentations with it or as precautions to avoid any risks from grading perspective. This is evident from the response of one of the students to the open-ended question related to application aspect: “*I use GenAI to create study plans and skill learning plans. I also use generative AI to create content for my resume and cover letter. Additionally, it also comes handy for generating simple emails too. However, it’s important to understand to crosscheck the authenticity of the information due to the hallucinating nature of AI for tools like ChatGPT,*

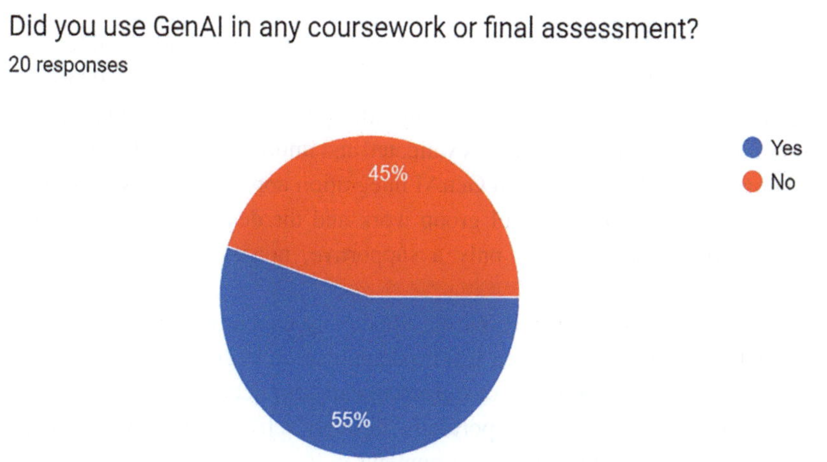

Fig. 5.4 Evaluation results for immediate survey for application aspect

DeepSeek, perplexity, etc. Along with this, I am well aware of the environments and impact of its usage in case of academic field hence I avoid its usage in the completion of final assignment."

Open-ended reflections reinforced these quantitative trends. Notable insights about specific task where GenAI was most helpful to the students during the course, regardless of whether they used it for the final assessment, include the following:

- *Ideas about problem related to the stakeholder management in real companies, to identify possible topic for the final assignment.*
- *Ideas about the video presentation structure.*
- *A specific task was to get a basic structure of the final assessment, which was to certain extent helpful but again required human intervention to standardize it as per expectations.*
- *We got good information about rural health innovations that was required for a group work in one of the sessions, but it was validated and improved from expertise of group members and secondary data.*
- *In the final exam, we have to find problem, simulate team environment, and provide solution, in an interactive video. Technology was useful to brainstorm, i.e., identifying some cases (of course require validation), asking for directions about problem, group dynamics, or solution.*
- *I have used it to gather the information and filter it according to my preference; for example, I have used it to gather information on Netflix and filter it, which helps me to reduce the time spending on different sites and websites for the same. Of course, the filter process requires data from other data sources to validate the information.*
- *I used it to elaborate some context which I had difficulty to understand and asked to provide me some examples and real cases and convert the difficult sentences into more meaningful versions.*

The analysis of the qualitative responses suggests that initial applications of GenAI were cautious and selective. Immediately after the intervention, many students described using GenAI for brainstorming or early-stage idea generation. The technology was also used for information generation but was validated and improved with information obtained from other sources. This aligns with GenAI integration approaches presented in Chap. 2, which emphasizes the challenges of group work and the design of applied, open-ended assignments where GenAI plays only a supportive, non-substantial role in assisting students.

This demonstrates that the intervention effectively encouraged hands-on exploration and collective experimentation, even if formal application in graded work was selectively adopted. As a result, the M&E goal of triggering early application was achieved—not just through submissions, but more importantly through active participation and practical exposure within structured learning environments.

5.1.4 Awareness: Ethics and Technology Landscape

This dimension examines whether students gained basic awareness of ethical issues in AI use and became familiar with other GenAI technologies beyond ChatGPT. Two questions, apart from the open question, captured awareness: one Likert-based (ethical understanding) and another binary (exposure to other tools).

Ethical Awareness (Likert)

Responses to the question "*How much basic awareness of ethical AI use you gained during your course?*" were as follows:

- 20% selected 5 (Strongly Agree).
- 45% selected 4.
- 20% selected 3.
- 15% selected 2.

Figure 5.5 shows that a total of 85% of students rated their ethical awareness at level 3 or above, indicating that the course effectively introduced them to the core ethical considerations surrounding GenAI use—such as responsible authorship, plagiarism, and data integrity. That 65% selected value 4 or 5 suggests that many students moved beyond passive awareness toward active reflection.

The 15% who selected value 2 may reflect students who require more structured opportunities to grapple with complex ethical dilemmas, possibly with good use cases. Another reason is that ethical issues are complex to master because technology itself is complex and requires careful judgment, critical thinking, and subject-matter knowledge. As students experiment more with technology and evaluate outcomes using their own insights, their awareness of ethical issues is likely to become more evident.

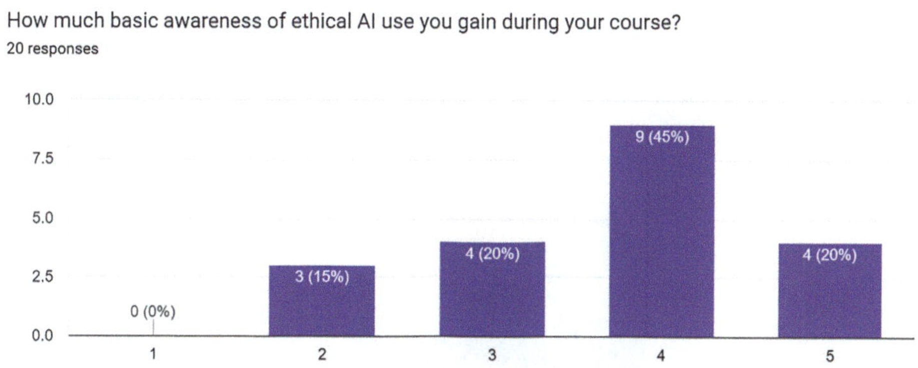

Fig. 5.5 Evaluation results for immediate survey for awareness aspect

Awareness of Other GenAI Technologies (Binary)

Students were also asked: *"Did you become aware of the existence of other GenAI technologies?" The survey results as graphically represented in* Fig. *5.6 indicate the following:*

- 63.2% answered yes.
- 36.8% answered no.

This result confirms that a majority of students (nearly two-thirds) became aware of alternative tools such as Microsoft Copilot and Gemini, even if they did not directly use them. This suggests that the intervention was successful in broadening students' conceptual exposure beyond a single platform, fostering an ecosystem-level awareness.

The open-ended responses about ethical concerns while using GenAI in the course further demonstrated that students had begun to consider ethical challenges:

- *We were unsure whether we had to disclose using GenAI to generate ideas (elementary use). However, we were instructed to present our findings using a two-color scheme (blue—ChatGPT output matches what we found through research/expert advice—and red—where ChatGPT output was wrong). Acknowledgement and reflection of technology become evident as part of group work presentation.*
- *Copy pasting AI-generated content as it is one of the major ethical concerns must only be used for knowledge collaboration but not on practical assignments.*
- *Plagiarism, biased outputs, privacy,* etc.
- *Wrong outputs, plagiarism, and acknowledgement to the technology use.*
- *Plagiarism, authenticity and originality, misinformation, and accuracy.*
- *I learned that GenAI can sometimes give wrong or biased information, so we have to check it carefully. We should also be careful not to share personal or private information with AI tools; it is important to avoid using AI to copy someone else's work or cheat in assignments.*

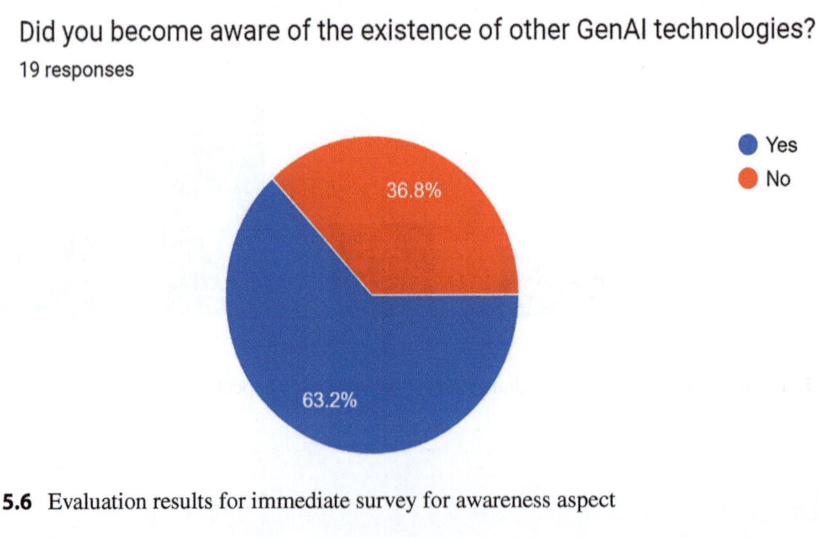

Fig. 5.6 Evaluation results for immediate survey for awareness aspect

Table 5.1 Summary of immediate evaluation results

Aspect	Key quantitative findings	Summary of findings
Knowledge	95% of students rated 3 or above; 65% rated 4 or 5	Most students developed a clear understanding of GenAI capabilities and limitations
Confidence	90% rated 3 or above on both confidence related survey questions; Likert value 3 was selected by 45% f or the first question and 40% for the second	Students feel confident using GenAI and are also confident explaining it to peers
Application	55% reported using GenAI in coursework or final assessment; actual engagement is likely higher. The remaining 45% reported not using GenAI in coursework or final assessments	As part of course redesign and delivery adaptations for responsible GenAI adoption, GenAI usage in in-class tasks and group-based work was common. A majority of students used GenAI in online learning system tasks or final graded assignments (55%)
Awareness	85% rated 3 or above on ethical awareness; 63.2% reported awareness of other GenAI tools	Strong initial awareness of ethics and broader GenAI ecosystem; some students began to reflect critically on ethical AI use

Finally, the analysis of qualitative responses suggests that students developed strong awareness of ethical and integrity-related issues. Immediately after the course, responses emphasized concerns about "copy pasting AI content" and the need to treat GenAI as a "knowledge collaborator" rather than a substitute. These reflections directly link to the GenAI integration approaches presented in Chap. 2, emphasizing the use of AI contribution forms, GenAI integration rubrics, GenAI usage instructions, and classroom discussions on technology functionality and limitations, including ethical issues involved.

Together for awareness aspect, the Likert and binary data, along with qualitative data, indicate that students recognized GenAI's ethical implications and also started considering these aspects during their experimentations with the technology.

To synthesize the results presented in this section, Table 5.1 summarizes the key findings across the four evaluation aspects—knowledge, confidence, application, and awareness—as measured immediately after the course. The table highlights both the quantitative trends and interpretive insights, reflecting how students initially responded to GenAI integration in academic contexts.

5.2 Follow-Up Evaluation Results: Analysis of Outcomes and Impacts

This section presents the results of the follow-up (medium-term) evaluation, conducted between 3 and 6 months after students completed the academic courses that integrated generative AI (GenAI) tools. This evaluation represents the second phase of the monitoring and evaluation (M&E) process and is focused on measuring medium-term outcomes that reflect knowledge retention, sustained application, and deeper awareness of GenAI technologies and ethics.

The objective of the follow-up evaluation, as outlined in the M&E framework, is to determine whether the skills, confidence, and awareness developed during the course were retained, expanded, and applied in new contexts over time. Specifically, the follow-up assessment examines the following:

- **Knowledge**: whether students have retained or expanded their understanding of GenAI and explored new concepts through self-study, practical use, or professional experiences.
- **Confidence**: whether students feel confident applying GenAI to more complex tasks beyond their coursework, and whether they share or teach GenAI concepts to others.
- **Application**: whether students continue to use GenAI in their academic, professional, or personal lives, and how they apply the tools in diverse scenarios such as thesis work, workplace tasks, or creative projects.
- **Awareness**: whether students have developed critical awareness of the ethical implications of AI, and whether they have experimented with other GenAI technologies beyond ChatGPT.

This phase is particularly important as it captures the longitudinal impact of the intervention and reflects whether GenAI literacy and responsible usage persisted longer possibly beyond the academic setting. By gathering insights after a temporal gap, the follow-up evaluation reveals whether the course served as a catalyst for continued learning, ethical reasoning, and real-world application.

As with the immediate evaluation, this section draws on data collected through a comprehensive survey tool including Likert-scale questions, binary responses, and open-text fields. The following subsections analyze the data across the four evaluation aspects, allowing for triangulated insights into how the GenAI intervention influenced students over time.

5.2.1 Knowledge: Retained and Expanded Understanding

This aspect assesses whether students have retained key concepts learned during the course and whether they have continued to expand their knowledge independently after completing the course.

The survey results indicated the following:

- 16.7% selected 5 (Strongly Agree).
- 50% selected 4 (Agree).
- 25% selected 3 (Moderate Agreement).
- 8.3% selected 2 (Disagree).

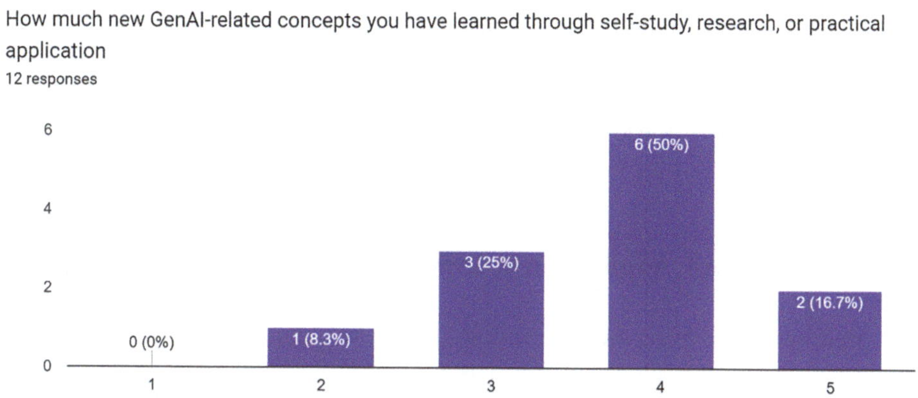

Fig. 5.7 Evaluation results for follow-up survey for knowledge aspect

Figure 5.7 shows that a total of 91.7% of students rated their knowledge at level 3 or above, suggesting that the vast majority retained meaningful understanding of GenAI and continued to engage with it beyond the course. Half of the students expressed confident agreement (value 4) that they had learned new concepts through self-study, research, or applied use.

The small proportion (8.3%) at value 2 may represent students who did not pursue GenAI exploration after the course maybe due to limited relevance to their academic or professional contexts. Nonetheless, the data reflects a strong medium-term outcome aligned with the M&E goal of knowledge reinforcement and expansion.

Open-ended reflections reinforced these quantitative trends. Notable insights about GenAI concepts learned by the students after the course include the following:

- *The role of GenAI in business ideation and helping design the survey questionnaire.*
- *How the companies are using GenAI, especially the educational institutions, from practice articles, especially from Harvard Inspiring Minds. The more you learn and the more experimentation you do, the better is the understanding.*
- *GenAI was helpful to me in navigating the job market (for example, by helping in creating a CV) and explaining some legal processes related to visa status and such. But I do have to gather insights from official websites and other information available publicly and advise from relevant experts.*
- *Multimodal modals, enterprise integration, and synthetic data generation.*
- *The course helped me understand the difference between helpful use and overreliance. I went further to develop skills in experimenting with the technology and learned how to ask for structured outputs.*

The analysis of qualitative responses suggests that students expanded this base knowledge through self-study and professional application. References to enterprise integration, multimodal tools, design of the survey questionnaire, and career-related uses (e.g., CV drafting) suggest that the structured introduction in the classroom created a foundation that encouraged continued exploration beyond the course. The experiences gained during the course and further self-exploration of the technology seem to have enhanced their knowledge levels.

5.2.2 Confidence: Application in Complex Contexts and Peer Sharing

This aspect evaluates two critical dimensions of students' confidence several months after the course, namely (1) their confidence in using GenAI for complex, real-world tasks that require more than just prompting and (2) their comfort with supporting peers in understanding or using GenAI. The survey indicates the following:

Confidence in Complex Tasks (Q1)

- 8.3% selected 5 (Strongly Agree).
- 58.3% selected 4 (Agree).
- 33.3% selected 3 (Moderate Agreement).

Confidence in Supporting Others (Q2)

- 0% selected 5.
- 33.3% selected 4.
- 50% selected 3.
- 16.7% selected 2.

Figure 5.8 shows that 100% of students rated their confidence at 3 or above, which is significant because this question targets not basic AI usage but strategic integration of GenAI into multifaceted academic or professional workflows.

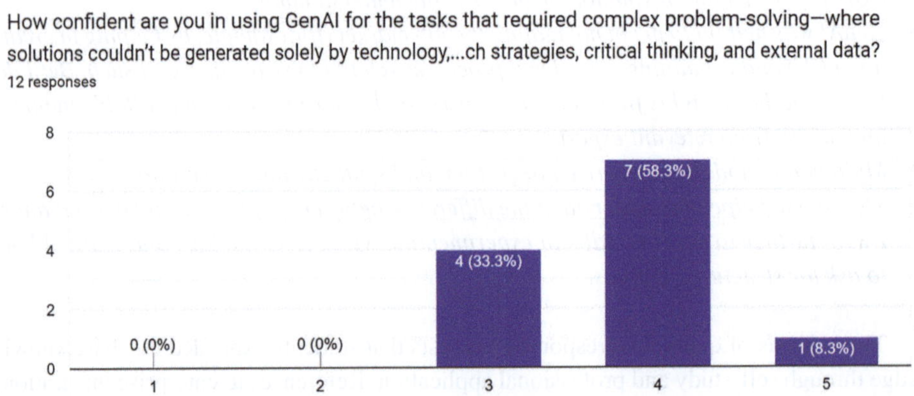

Fig. 5.8 Evaluation results for follow-up survey for confidence in using technology for complex tasks

Higher values indicate that students are more confident in using technology for more complex tasks owing to their good understanding of the functionality and limitations of the technology, at least based on their own experiences accumulated over a prolonged period, starting from the first structured experiments conducted during the course. This means that majority of the students feel confident in using this technology as a knowledge collaborator, taking its outputs as directions, and finding final solutions for the complex problems driven by facts brought by extensive research, critical thinking, and data available elsewhere. In other words, this indicates that students are not only competent in operating GenAI tools but also increasingly confident in using them as cognitive collaborators—to support problem-solving, synthesize sources, refine arguments, and generate solutions requiring both AI output and human judgment. Such results represent a strong fulfillment of the M&E framework's medium-term outcome: students developing sustained, high-level confidence in working alongside GenAI to enhance intellectual tasks—not replace them.

Regarding the confidence of students in supporting colleagues, classmates, or friends with the technology, Fig. 5.9 shows that while no student selected the highest confidence level, 83.3% rated themselves at 3 or above.

This suggests that most students have engaged in some form of peer support—whether by sharing tools, explaining basic uses, or offering prompt-writing tips. The presence of moderate confidence rather than high self-rating reflects realistic self-assessment: students recognize that they are still learners, but they feel capable of helping others get started, especially with foundational or ethical aspects of GenAI use. Open-ended reflections about confidence in using or sharing GenAI-related concepts at a level beyond basic understanding reinforced these quantitative trends. Notable insights include the following:

- *I am experimenting but remain cautious. It is easy to accept what GenAI says, even when it seems incorrect. The more you work with the technology, the more confident you become, yet its potential use cases are vast and open-ended. However, the*

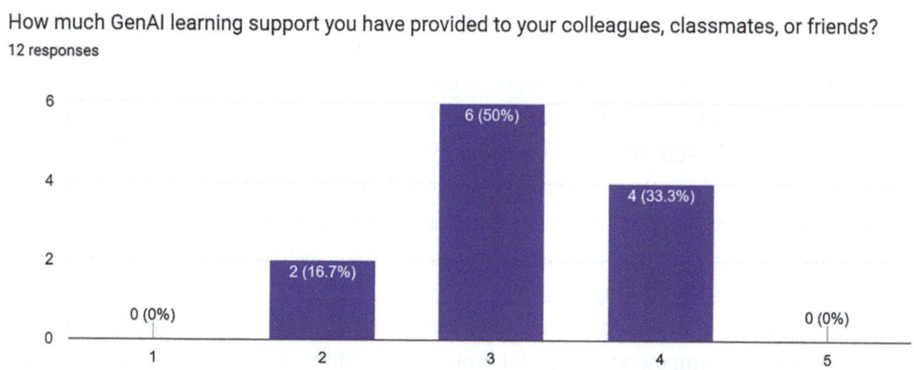

Fig. 5.9 Evaluation results for follow-up survey for confidence in supporting peers with technology-related aspects

confidence gained from experimenting and explaining the technology makes the exploration collaborative and enjoyable, which further reinforces confidence levels.

- *I am very confident but provided only minimal support. This may be why some people are still hesitant to share that they want to use—or are already using—GenAI in their work. Sometimes people are hesitant to share that they are using GenAI because not everyone views it positively.*
- *Helped peer to use the technology document upload functionality, providing copyright-related information (what is allowed and what not). This was driven by our group work where we were instructed about technology ability to analyze larger files and associated limitations of the technology.*
- *The technology has numerous use cases. I am confident to explain what I found, but for other use cases, I need to explore it further. But the experimentation skills are important that could help leverage existing experience with technology in providing support to the peers for their needs.*
- *I was able to explain to one of my peers how to summarize the research paper and then validate the findings based on the outcome of other technologies like Elicit and own analysis of the research paper. I was confident to explain based on my understanding of the technology functionalities and limitations (point where it will give wrong responses).*
- *I can explain my learning with this technology for the tasks that I completed with its support.*
- *I am confident in explaining concepts that I explored. Recently, I explained to my peer how to use GenAI in documentation writing. You can get nice ideas about structure, improvement of expressions, and text generation of the sections (based on the context provided to the technology) that otherwise will require lot of work to draft the section that is merely introductory.*
- *I am able to explain to them how I used GenAI as a knowledge collaborator, for purposes like overcoming progress blockage or to have a better understanding of certain concepts to help me continue my academic work.*

The analysis of qualitative responses suggests confidence translated into a readiness to independently explore complex applications and even guide others. Some students described sharing experiences with peers outside the classroom, showing that the initial scaffolding provided by structured activities, AI policies, and contribution forms enabled sustainable growth in self-efficacy. Additionally, one response suggests that GenAI users may be reluctant to share their expertise with peers, as some hesitate to disclose their use of GenAI due to mixed perceptions—while some view it as a valuable learning aid, others regard it with skepticism, associating it with factors like overreliance, academic dishonesty, or reduced critical thinking.

Taken together, confidence dimensions demonstrate that students are progressing beyond individual competence to collaborative application and peer leadership, which reflects strong developmental growth since the initial exposure during the course.

5.2.3 Application: Continued and Contextual Use of GenAI

This aspect captures whether students continue to use GenAI after the course and whether they have adapted it to new settings, such as research, professional tasks, or personal projects. The survey results indicate the following:

- 100% of students responded Yes.

Figure 5.10 shows a unanimous affirmation of continued use, providing clear evidence that students have internalized GenAI as a functional tool in their academic, professional, or personal workflows.

Open-ended responses further revealed how students used GenAI in other tasks like research, professional projects, personal projects, and other courses. This includes the following:

- *There are some complex statements in industry case studies (case reports). The technology made it easier to understand those complex terms.*
- *Improving the presentation expressions as I do not come from native English-speaking country. But you need to be good to understand if the GenAI-generated expressions are equivalent to the one you wish to alter.*
- *Used GenAI to summarize academic papers and explain terms in plain language.*
- *Edited my short research report and got instant feedback on tone and clarity.*
- *I used it to summarize research paper (open access) and some difficult terminologies of the paper.*

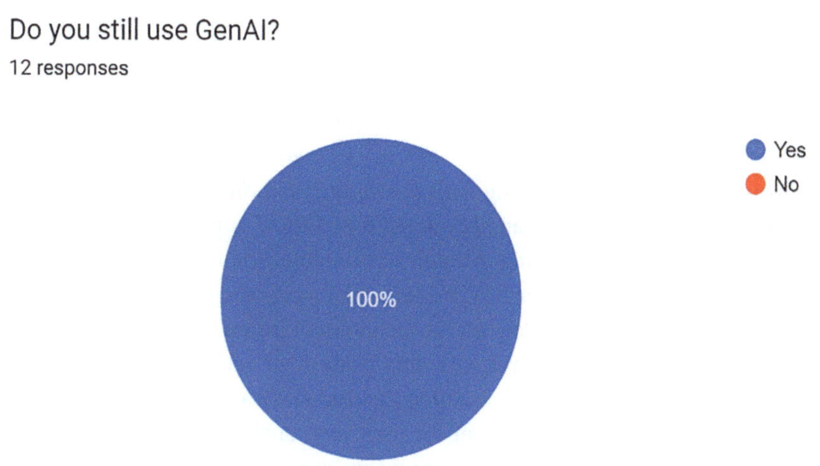

Fig. 5.10 Evaluation results for follow-up survey for technology application aspect

- *I used it to design the survey questionnaire using indicators that derived from literature review, along with the inputs from different research articles, for instance direct questions.*
- *Explored some topics for my thesis and organizing workflows.*
- *I used it to identify the areas of improvement of my CV for particular job application and also improved my LinkedIn profile. Also used it to check tone in professional emails (culturally sensitive).*
- *GenAI was helpful in suggesting a potential breakdown (or table of content) in other course assignments; most of the time, the suggested titles and subtitles were relevant and accurate. For work purposes, GenAI was helpful in tailoring my CV to fit the German CV format to have better chances in job applications. It was also helpful when learning German, where it explained grammar or linguistic concepts effectively.*
- *Yes, after completing the course, I used GenAI to support a personal project involving the creation of a small business plan. I relied on GenAI to:*
  - *Generate initial ideas for the business model and value proposition.*
  - *Refine my writing for clarity and tone in the executive summary.*
  - *Create visuals such as mock-up logos and basic infographics using multimodal tools.*
  - *Analyze market trends by summarizing online articles and synthesizing key points.*
  - *Simulate investor Q&A, helping me prepare for presenting the idea to others.*
  - *GenAI significantly sped up the process and improved the quality of my outputs. It allowed me to move from ideation to a polished pitch deck in a fraction of the time it would normally take.*
- *All routine and time-consuming tasks like summarizing my work and drafting emails, social media contents, emails,* etc. *It is easier to adapt the GenAI outcome rather than drafting them from scratch.*

The analysis of qualitative responses suggests the continued use of GenAI, i.e., all students reported integrating it into their academic, professional, or personal tasks. The qualitative comments show that usage became more purposeful and nuanced over time. One student explained, "*I used it to design the survey questionnaire using indicators that derived from literature review, along with the inputs from different research articles, for instance direct questions*," while another noted how GenAI supported them in summarizing research papers and explaining difficult terminologies. Others highlighted its use in creating LinkedIn content, German CV, business plan portions, etc. The responses indicate that students are not fully dependent on the technology; instead, they use it selectively for specific activities, complementing it with data from other sources and their own manual efforts. The application pattern suggests that the students framed GenAI as a "knowledge collaborator" rather than a replacement for their own analysis. Such reflections demonstrate the impact of GenAI integration approaches explained in Chap. 2, particularly the deliberate design of assessments that could not be solved by AI alone, the use of AI contribution forms, and the requirement for explicit reflection on AI usage. These mechanisms appear to have instilled habits of critical application—students learned to position GenAI as an assistant for ideation, structuring, and exploratory analysis while maintaining human oversight and responsibility for outcomes.

5.2.4 Awareness: Ethical Maturity and Technology Exploration

This aspect explores whether students have developed critical awareness about AI ethics and whether they are actively experimenting with new tools beyond ChatGPT. The survey results indicate the following:

Encouraging Responsible Use (Q1)

- 100% responded Yes.

Use of Other GenAI Technologies (Q2)

- 41.7% responded Yes.
- 58.3% responded No.

Figure 5.11 shows a full commitment among students to responsible GenAI advocacy, demonstrating a high level of ethical maturity. This suggests that the intervention helped students not only use AI ethically but also influence their peers, which is a vital medium-term impact. This requires students to critically reflect on whether the use of technology is categorized solely as a collaborator with no eligibility for the authorship with acknowledgement and clear disclosures of where it is used (ethical) or the other way around (unethical). As per one of the students, "*A friend of mine used GenAI to help draft content for their assignments. Initially, they used it for brainstorming and idea organization but eventually started relying on it to generate full paragraphs (although the paragraphs were*

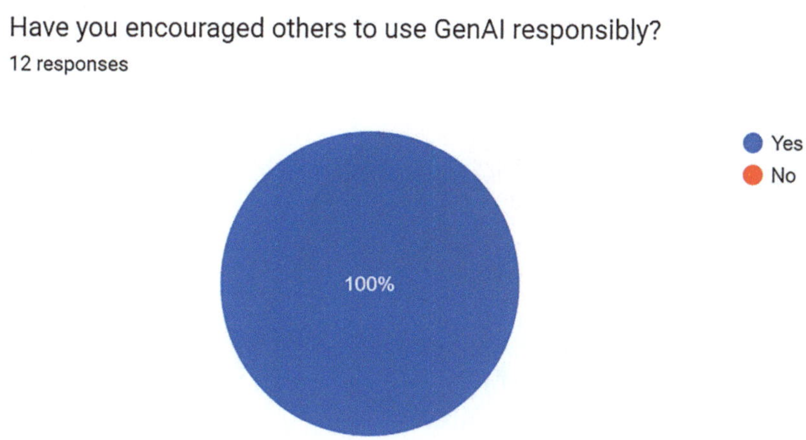

Fig. 5.11 Evaluation results for follow-up survey for students' awareness of responsible technology use

not the part of the solution). The ethical dilemma arose when students questioned whether submitting AI-assisted work—even when the AI contribution was limited to contextual support rather than the core solution—truly reflected their own learning and effort. They were aware of issues related to knowledge collaboration and authorship when considering the assignment as a whole. They felt undecided between using a helpful tool and maintaining academic integrity. Ultimately, they decided to disclose the use of AI and only use it for planning and feedback—not final writing—to stay within ethical guidelines. Later, they were very aware of ethical issues and in future research papers they first check the publisher GenAI guidelines and then decide whether or not to use technology even as a knowledge collaborator." This represents that awareness they got in the courses had been expanded beyond the course and applied in future projects.

Meanwhile, 41.7% of students experimented with other tools—such as Microsoft Copilot—indicating a budding curiosity and expansion beyond ChatGPT (Fig. 5.12). While a majority still relied on ChatGPT alone, the data still reflects growing ecosystem awareness, especially for nontechnical users.

Open-ended responses further revealed how students having good awareness of ethical issues with GenAI technologies faced ethical dilemmas. Notable insights include the following:

- *I along with my groupmate faced dilemma between using technology for simplifying the work* vs. *its reporting (ensuring its usage is clearly acknowledged). This made us think how much contribution it could have and in what roles. The AI contribution assessment form and GenAI integration rubrics of the professor were really helpful, but publisher policies need to be considered.*

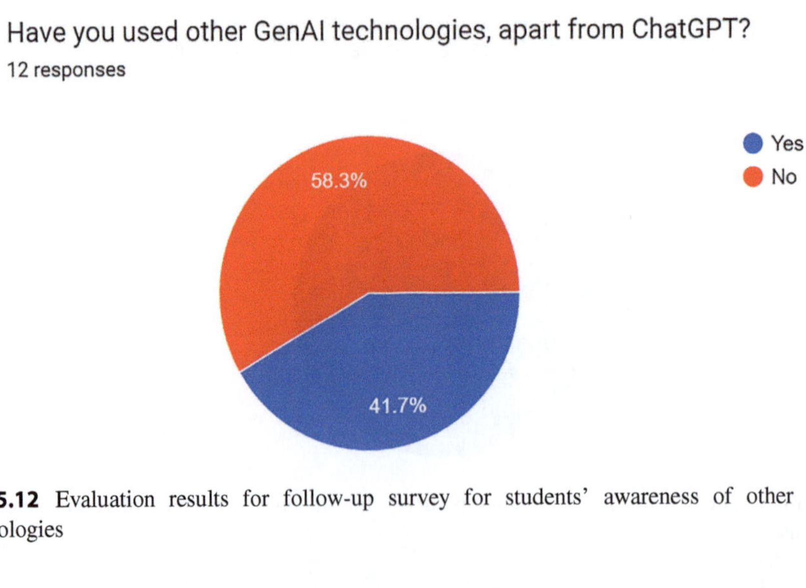

Fig. 5.12 Evaluation results for follow-up survey for students' awareness of other GenAI technologies

- *It is always about acknowledging the use of GenAI. When you do this, some people lack trust in your work.*
- *We cannot upload our own research paper draft to technology as we were not sure if it will get the data for training. On the other side, we need to get insights about areas where language is to be corrected. After correction, issue was to acknowledge the support of technology, and publisher requirements were to be checked about technology use.*
- *We discussed whether GenAI assistance should be acknowledged in professional presentations, although the use is limited to seeking directions. It is sometimes harder to explain those directions but AI contribution assessment form is useful to report the AI usage.*
- *In survey questionnaire, there were some questions (pertaining to the indicator) that made lot of sense, but was not sure if it comes from existing literature or created afresh by the technology. So, whether we acknowledge the technology or cite the article. The reference generated by the technology does not exist at all, but the survey question was valid.*
- *I read about AI-written articles, with higher plagiarism, and that concerned me. This is good lesson as even using GenAI content, you can have plagiarism, even when you never copied from other sources.*
- *In one of my assignments, ChatGPT generated information about a company that I could not verify through the company's website or recent annual reports. It was difficult to decide whether to include that information or not, but eventually I left it out because I was not able to cite an actual source that supports it.*
- *Even we can use business version of technology or revoke our consent to use our shared content for training, yet we are still not sure if it really happens. It is better to avoid it for sensitive tasks.*

The analysis of the qualitative responses shows that awareness extended beyond classroom ethics into professional and personal domains. Students described actively checking publication or institutional policies before using AI in research, reflecting concerns about trust and credibility. One remarked, "*It is always about acknowledging the use of GenAI. When you do this, some people lack trust in your work*," pointing to dilemmas around transparency and audience perception. They struggle with how to properly acknowledge some AI contributions, for instance using tool for seeking directions, while maintaining trust and credibility, and they find the AI contribution assessment form and integration rubrics helpful for guidance. Concerns also arise around data security and publisher policies, particularly when uploading drafts or using AI for sensitive tasks. Students noted difficulties in verifying AI-generated content, such as company information, and recognized that even AI-assisted work can unintentionally lead to plagiarism. Additionally, the valid information generated by the GenAI, for instance valid survey question, could be harder to be acknowledged as the supporting references are fake and will not exist at all. Overall, while AI is seen as a useful tool, students emphasize the need for careful judgment, transparency, and adherence to ethical and procedural standards.

Table 5.2 Summary of follow-up evaluation results

Aspect	Key quantitative findings	Summary of findings
Knowledge	91.7% rated 3 or above; 66.7% rated 4 or 5	Strong retention and expansion of GenAI knowledge through independent use, research, practical application, and experimentations with it after the course
Confidence	100% rated 3 or above for complex problem-solving; 83.3% rated 3 or above for peer support	Students feel confident using GenAI strategically; most are also able to guide others beyond basic understanding
Application	100% reported continued GenAI use	Universal post-course adoption, with use cases spanning academic, professional, and personal domains
Awareness	100% encouraged responsible use; 41.7% tried other tools	Strong ethical engagement; partial exploration of the broader GenAI ecosystem beyond ChatGPT

These medium evaluation results reinforce that students are now active participants in the evolving GenAI landscape, equipped with both critical thinking and ethical grounding. Several responses noted risks of plagiarism, biased outputs, and misinformation, reinforcing that critical evaluation of AI outputs had become a routine part of their practice. This aligns with Chap. 2's strong emphasis on ethics through in-class discussions, explicit rubrics, and group presentation strategies that required students to highlight correct vs. incorrect AI-generated content. By embedding structured ethical reflection into assignments and group work, the intervention appears to have fostered knowledge, confidence, application, and awareness that persisted and evolved. Students were not only applying ethical guidelines themselves but also advocating responsible use among peers, echoing the course's intent to normalize transparency and reflective practice in AI adoption.

Table 5.2 tabulates the summary of follow-up evaluation results.

5.2.5 Synthesis of Evaluation Findings

The evaluation of GenAI integration in academic courses was conducted in two distinct phases: an immediate evaluation survey at the conclusion of the course and a follow-up evaluation survey several months later. Although the immediate and follow-up evaluation surveys were completed by different student cohorts, both groups participated in similarly designed courses with comparable instructional guidance, technologies, and assignments. As such, the results can be viewed as complementary snapshots of how students at different stages engaged with GenAI within the same pedagogical framework. The two-tiered approach, rooted in the monitoring and evaluation (M&E) framework presented in Chap. 3, allowed for a better understanding of both the short-term learning outcomes and the medium-term developmental impact of the intervention and fostered the evaluation process.

The results of both surveys demonstrate a consistent and meaningful impact across all four outcome aspects—knowledge, confidence, application, and awareness—providing strong evidence that the intervention succeeded in building GenAI-related competencies that extend well beyond the classroom.

- **Knowledge**
  Immediately after the course, 95% of students rated their understanding of GenAI concepts at level 3 or above, with 65% at level 4 or 5. In the follow-up evaluation, this trend persisted and strengthened: 91.7% still rated 3 or above, with two-thirds selecting 4 or 5. Students across both groups reported strong knowledge levels, with the follow-up group demonstrating signs of further engagement and self-directed learning.
- **Confidence**
  In the immediate evaluation, students reported generally positive confidence in both using GenAI and explaining it to peers. While 90% rated themselves at level 3 or above on both confidence-related questions, 45% selected value 3 for confidence in using GenAI and 40% for confidence in explaining it to peers. The follow-up group showed strong confidence in applying GenAI to complex tasks, possibly reflecting longer exposure or real-world testing of concepts initially introduced in class. All participants (100%) in the follow-up evaluation rated themselves at level 3 or above in their ability to use GenAI for complex, multilayered tasks requiring critical thinking, research, and human judgment. A majority also reported confidence in supporting peers, indicating a shift from individual competence toward collaborative confidence, with 83.3% rating themselves at level 3 or above
- **Application**
  The difference between the two surveys was most striking in the area of application. In the immediate survey, 55% of students reported using GenAI in formal coursework or final assignments. However, further analysis revealed that many of those who answered "No" had in fact used GenAI during class sessions or for group-based learning but chose not to use it in graded submissions—possibly due to concerns about academic misconduct or uncertainty around institutional policies. Immediate group began with in-class and moderate formal application, while the follow-up group (later in their academic or professional paths) reported full adoption, suggesting that the model promotes real-world application when given time. For instance, all respondents of follow-up survey (100%) confirmed continued use of GenAI, demonstrating that students had integrated the tool into their academic, professional, and personal routines. This complete adoption reflects the course's success in equipping students with not just the tools, but the confidence and ethical grounding to use GenAI meaningfully and autonomously.

- **Awareness**
  Awareness of both ethical considerations and the broader GenAI ecosystem are different for both immediate and follow-up surveys as expected. Immediate survey respondents (85%) indicated some level of ethical awareness, and 63.2% were aware of other GenAI tools. However, all follow-up survey respondents (100%) reported encouraging responsible GenAI use among peers—showing a clear internalization of ethical principles. While fewer students (41.7%) had explored tools beyond ChatGPT, the shift toward critical awareness and cautious experimentation was evident.

6 Recommendations and Future Directions

The findings and experiences presented throughout this book underline one central message: generative AI (GenAI) is neither a threat to academic integrity nor a solution to student learning challenges. Rather, it should be framed as a knowledge collaborator with non-substantial contributions that enhance learning when integrated responsibly into the design of teaching, learning activities, assessments, and evaluation frameworks. Importantly, these integrations must be accompanied by systematic monitoring and evaluation (M&E) to ensure that they deliver meaningful, measurable impacts, both in short and long terms. In this chapter, the authors bring these threads together, offering practical recommendations and future directions for educators, institutions, and policymakers who wish to harness GenAI responsibly and effectively.

6.1 Embedding GenAI Across Course Activities

GenAI should be integrated across the entire course, rather confined to graded assignments. It can enrich group projects, in-class discussions, research activities, and reflective exercises. For instance, when discussing results from real consulting projects, students might be asked to use both GenAI and their own research to improve upon those outcomes that are results of the real project. This approach does not replace student effort but challenges them to compare machine-generated outputs with human reasoning and real results, highlight gaps, and propose refinements. In such a model, experimentation with GenAI becomes part of everyday learning, helping students appreciate its strengths and limitations while developing the skills to integrate it ethically into complex problem-solving. Short training programs, including self-paced modules, could be offered to students before the start of the course. These would familiarize them with the working, functionalities, and

V. Gupta, C. Gupta, *Technological Innovations in Education*, Synthesis Lectures on Technology Management & Entrepreneurship,
https://doi.org/10.1007/978-3-032-11225-5_6

limitations of GenAI, thereby enabling them to apply these concepts more effectively and elaborate on them in practice during the redesigned course.

6.2 Redesigning Assessments and Rubrics

Traditional assessments reliant on knowledge recall are increasingly insufficient in the age of AI. Instead, assignments should emphasize applied, open-ended, and context-driven problems that demand creativity, justification of choices, and practical feasibility. These are tasks that GenAI cannot resolve independently. To support transparency, students should use instruments such as the GenAI integration rubric, AI contribution form, and AI contribution assessment form. These documents guide students in reflecting on their own technology use, distinguishing between substantial and non-substantial contributions, and openly acknowledging AI assistance. When integrated into group work, such forms also help teams negotiate ethical boundaries and share responsibility for technology use. The incremental experiences with these forms and technology use under structured safeguards during the course will be valuable during graded assignment research work.

6.3 Establishing Clear and Flexible AI Policies

Every course should communicate its AI policy up front. This policy must explain what role GenAI is allowed to play, why those boundaries exist, technology functionality and limitations, and how students are expected to report and acknowledge its use. Short training sessions on how to complete AI usage forms and interpret the rubrics and continuous use of these documents during the entire course can foster the process. At the same time, policies must remain flexible, tailored to the course's learning objectives, nature of the courses, and tutors' decision whether to integrate GenAI or not in the course. For example, a consulting course may actively integrate GenAI into both group tasks and assessments, while a clinical psychology practicum may prohibit it altogether due to ethical and confidentiality constraints.

6.4 Customization of Supporting Documents

The documents introduced in this book—such as the GenAI integration rubric, AI contribution form, AI contribution assessment form, GenAI usage instructions—should be viewed as living templates. They are intentionally flexible, meant to be adapted to specific course objectives, assignment formats, and disciplinary needs. Customization helps students see their direct relevance that they could apply in their research works, making it easier to connect rubric guidance with real learning tasks and to understand what constitutes appropriate, non-substantial use of technology, and how to acknowledge the technology usage. Over time, building discipline-specific versions of these documents can foster stronger alignment between ethical AI use and professional practices.

6.5 Leveraging Custom GPTs for Learning Support

Future teaching can also be enhanced by the creation of custom GPTs that are tailored to course content, AI policy, and AI documents required for understanding technology's possible contributions and reporting its use. These tools can help students navigate complex concepts, clarify AI policies, or interpret documents such as GenAI integration rubrics and AI contribution form. By embedding such support directly into the learning environment, tutors can reduce administrative overhead while giving students an accessible, interactive resource. There are numerous resources available for creating course-specific custom GPTs. For instance, the tutorial (https://hbsp.harvard.edu/inspiring-minds/create-custom-gpts-for-your-course) provides step-by-step guidance on designing custom GPTs. The custom GPT has already been successfully applied in marketing analytics and data mining courses at Elon University, USA.

6.6 From Pilot Experiments to Scalable Practices

One of the most important lessons from GenAI integration so far is that success often begins small. Pilots conducted in individual courses allow tutors to experiment with new forms of group work, rubrics, or evaluation instruments under controlled conditions. In other words, the tutors are always learning through their GenAI integration experiences and innovating the course AI policy and other integration approaches. When these pilots are paired with rigorous M&E—using the reusable surveys, contribution forms, and pre-testing protocols provided in this book—the results generate concrete evidence of what works. This evidence then becomes the foundation for scaling. A practice that proves effective in one course can be adapted across several modules within a program and eventually inform institution-wide strategies. Scaling in this way ensures that expansion is not based on assumptions or hype, but on validated outcomes and student feedback. In the long term, gradual scaling supported by evaluation also helps institutions build confidence, align practices across faculties, and prepare the ground for coherent GenAI policies at the university level.

6.7 Fostering a Culture of Responsible Experimentation

Ultimately, the goal is to create a culture where students can experiment responsibly with GenAI under structured safeguards. GenAI is innovating at a rapid pace, with new technologies emerging in the markets. Technology adoption decisions and associated digital transformations require users to experiment with new technologies, identify their use cases, and then make rational adoption decisions (Gupta, 2025). Reflection and transparency are key for structural experimentation outcome reporting for peer learning and making technology adoption decisions at larger scale in the organizations. By experimenting under structured safeguards, and using rubrics, AI forms, and acknowledgement practices, students develop habits of responsible use and ethical reporting that extend into professional life. They learn to engage in reflective synthesis—deciding when to draw on AI outputs, when to validate against other sources, and when to rely on their own judgment. Such habits are crucial not only for academic integrity but also for future workplaces where AI tools will increasingly be part of everyday practice.

6.8 Integrating Monitoring and Evaluation (M&E)

A distinctive contribution of this book is the full monitoring and evaluation (M&E) framework that accompanies the course interventions. The experiences discussed show clearly that integrating GenAI is not sufficient on its own; it must be monitored, evaluated, and refined through systematic processes. In practice, this means that every experiment with GenAI in the classroom during the course should be designed together with a theory of change (ToC) that clarifies how activities are expected to generate outputs, outcomes, and long-term impacts. Such a ToC ensures that evaluation is not an afterthought, but an integral part of course redesign and GenAI integration.

The framework presented here is practical and adaptable. Educators can reuse the ready-made instruments included in the book—such as the immediate and follow-up survey tools, pretesting protocols, and evaluation protocols—or adapt them to their unique contexts. These instruments already align with the four key outcome dimensions of knowledge, confidence, application, and awareness, meaning peers can adopt them directly without having to build new measures from scratch. Alternatively, new indicators aligned with the ToC could be easily added to the research instruments. The pretested survey questions, for example, are designed to capture both quantitative and quantitative aspects of impact of GenAI in short and long terms. By reusing these instruments, faculty can quickly begin collecting valid and comparable data while saving time and effort.

Just as important, process data such as the AI contribution form and the GenAI integration rubric provide a structured way to document how students actually used technology during their assignments. These artifacts are not only tools for transparency in assessment; they also aid in evaluation process.

The evaluation procedures outlined in the book also demonstrate how peers can structure their own cycles of learning. Immediate evaluations at the end of a course capture short-term changes across knowledge, confidence, awareness, and application, while follow-up evaluations several months later reveal whether students continue to apply responsible GenAI practices in professional or personal contexts. By replicating these procedures, educators can build an evidence base that extends beyond classroom performance, showing whether interventions create lasting behavioral change.

Finally, the M&E framework is designed for iterative refinement. Findings from surveys, contribution forms, and reflective assignments can be fed directly back into course redesign—informing updates to rubrics, adjustments to AI policies, or shifts in assignment design. In this way, M&E is not simply about accountability; it becomes a continuous improvement cycle. By reusing and adapting the procedures and instruments provided in this book, peers can embed evaluation into their GenAI practices from the outset without involving third-party evaluators, ensuring that integration is both responsible and evidence driven.

6.9 Sharing Results of Course-Specific Monitoring and Evaluation (M&E) Results

The GenAI integration across courses can be seen as a series of experimentations, with lessons learned through continuous monitoring and evaluation. These insights will be instrumental not only in improving GenAI integration practices within the courses where it was directly employed but also in other courses across the university. The results and lessons learned can be systematically documented using an experimentation outcome template, as proposed by Gupta and Nyamapfene (2025). This template requires faculty to briefly tabulate how GenAI was used in the course (for example, in graded assignments, group work, or classroom activities), what outputs and outcomes were achieved, what lessons were learned, and how these lessons were subsequently incorporated to further improve GenAI practices.

The idea is that faculty can use existing M&E frameworks, or adapt them, to conduct their own evaluations and prepare an impact evaluation report. These evaluations are formative, providing evidence and lessons that, over time, will strengthen summative evaluations of GenAI integration across the university, potentially conducted by third parties. An evaluation report should detail how the evaluation was carried out, the results obtained, and recommendations for improving GenAI integration. Such reports could be publicly shared, enabling the wider academic community to replicate the evaluations in their own courses and to implement successful GenAI practices—either directly or with contextual adaptations. This process also sets the stage for iterative evaluation and continuous improvement of practices.

The experimentation outcome template serves as a concise record, outlining the GenAI practices employed in a course, their M&E results, and the improvements planned for future iterations. Table 6.1 tabulates the experimentation outcome template.

Table 6.1 Experimentation outcome template *(taken from* Gupta and Nyamapfene (2025), *reused with the permission of the publisher; the figure remains under the standard Taylor & Francis license)*

Section	Description
Activity description where GenAI was used (for instance, assignments)	Provide a brief description of the assignment, including the nature of the problem students are required to solve as part of the assessment
Category of assessment	Category 2 *(assistive role)* or 3 *(integral role)* as per UCL GenAI guidelines
GenAI usage	Explain how GenAI is supposed to be used by students, e.g., for data analysis and formal analysis. Link back to the *GenAI integration rubric* (Table 6.1)
Expected outputs *(evaluation indicators)*	Describe what outputs are expected from the successful adoption of GenAI by the students. This should be linked back to theory of change (ToC)
Expected outcomes *(evaluation indicators)*	List expected short-term and medium-term outcomes *(depending on the type of evaluation)* aligned with theory of change (ToC)
Challenges/issues during GenAI integration	Identify any challenges or issues encountered during the experiment, e.g., difficulty in maintaining academic integrity, students struggling with ethical AI use, usage of AI humanizing technologies, or any requirement of instructional material
Recommendations for peers	Provide insights and suggestions for peers who might want to integrate GenAI into their courses. Focus on best practices, lessons learned, etc. Also mention to what type of courses the recommendations will apply and what courses will find it hard to apply them
Follow-up actions	State what follow-up actions are planned, such as modifying assignments, conducting further experiments, course redesign, and new artifacts
Survey information	Mention which indicators are accessed through surveys, and briefly how survey was conducted. Also mention which indicators were accessed through course documents, for instance AI contribution forms that aided the evaluation process. If possible, provide link to survey questionnaires for reuse possibility and link to aggregated data *(removing personal data)* and the evaluation report

6.10 Future Research and Policy Directions

Looking ahead, several areas require sustained attention. First, there is a clear need for longitudinal research that tracks how responsible GenAI use affects not only academic integrity but also professional practice and lifelong learning *(long-term impact)*. Capturing these long-term trajectories will help institutions understand whether classroom interventions translate into long-term impacts of GenAI.

Second, there is an urgent case for cross-institutional collaboration. Universities should not operate in isolation but actively share their evaluated experiences with GenAI integration. When successful cases are documented through rigorous monitoring and evaluation (M&E)—using instruments such as those provided in this book—they become powerful exemplars. Disseminating these evaluated cases enables peers to replicate or adapt effective practices, fosters a culture of responsible experimentation, and accelerates the refinement of integration strategies across different contexts.

Ultimately, sharing such evidence can enrich the body of knowledge on GenAI integration across courses, fostering a deeper understanding that may inform the development of institution-wide policies and, over time, contribute to sector-wide frameworks. At present, GenAI integration in academia appears to be more guidance rather than policies. By building a repository of evaluated interventions, including their ToC models, survey results, and reflective insights, universities can move toward a more coherent, evidence-based framework for GenAI in higher education.

References

Gupta, V. (2025). From experimentation to insight: Structured reporting of AI experimentations in libraries. *Journal of Web Librarianship, 19*(1), 93–104. https://doi.org/10.1080/19322909.2025.2450539

Gupta, V., & Nyamapfene, A. (2025). Integrating generative AI in academic assessments: Practical recommendations on AI usage and enhancing knowledge sharing. *Internet Reference Services Quarterly, 29*, 1–11. https://doi.org/10.1080/10875301.2025.2456903

Appendixes

Appendix I: Questionnaire (For Pretesting of Immediate Evaluation Survey Instrument)

Introduction

This pretesting questionnaire is designed to evaluate the clarity, logical flow, appropriateness, and other aspects of the immediate survey before its full-scale administration. The objective is to identify any problems with the survey in the pretesting phase rather than at the actual evaluation phase of the program. The immediate survey will be conducted later (after the pretesting) to assess the short-term impact of the intervention (GenAI integration in academic courses) on participants' knowledge, confidence, application, and awareness. Pretesting helps identify any issues with question wording, response options, or survey structure to ensure a smooth and effective data collection process of the intervention evaluation.

Before proceeding, please read the informed consent statement. Your participation is voluntary, and your responses will remain confidential. Your information like university emails will not be shared with anyone and will be deleted after pretesting report is completed. The final report will be published without identifying details, but with aggregated responses, with select quotes included after removing identifying details *(if quotes are shared by the survey participants)*. Individual responses will be retained for **one month** post-pretesting and then securely deleted. By continuing, you agree to participate in this pretesting phase.

V. Gupta, C. Gupta, *Technological Innovations in Education*, Synthesis Lectures on Technology Management & Entrepreneurship,
https://doi.org/10.1007/978-3-032-11225-5

Useful Instructions

The questionnaire is divided into five sections; first four sections represent one of the evaluation aspects: *knowledge, confidence, application, and awareness*. These sections contain the following:

- Evaluation questions designed to measure the short-term impact of the program on that aspect.

Last section represents the questionnaire-related aspects and contains the following questions:

- Pretesting questions to assess potential issues with the survey instrument, such as clarity, logical flow, appropriateness of response options, impartiality, and time required to complete the survey.

The evaluation questions and pretesting questions include 5-point Likert scale items *(1 representing strong disagreement and 5 representing strong agreement)*, binary *(yes/no)* questions, and open-ended questions to gather detailed feedback. If you find any question unclear, difficult to answer, or biased, please provide details in the open-ended sections. Additionally, a follow-up interview may be conducted to further understand your responses and gather any additional feedback you wish to share. If you agree to take part in the follow-up, please mention your university email at the end of the questionnaire; otherwise, leave it blank.

Knowledge

- **How well do you understand GenAI after completing the course?**
  - ☐ 1
  - ☐ 2
  - ☐ 3
  - ☐ 4
  - ☐ 5

Confidence

- **How confident are you in using GenAI for academic work?**
  - ☐ 1
  - ☐ 2
  - ☐ 3
  - ☐ 4
  - ☐ 5
- **How confident are you in explaining GenAI to the peers?**
  - ☐ 1
  - ☐ 2
  - ☐ 3
  - ☐ 4
  - ☐ 5

Application

- **Did you use GenAI in your final assessment?**
  - ☐ Yes
  - ☐ No

Awareness

- **How much basic awareness of ethical AI use you gained during your course?**
  - ☐ 1
  - ☐ 2
  - ☐ 3
  - ☐ 4
  - ☐ 5
- **Did you become aware of the existence of other GenAI technologies?**
  - ☐ Yes
  - ☐ No

Questionnaire Aspects for Pretesting

Clarity and Comprehension

- **Did you clearly understand all the questions in this survey?**
  - ☐ Yes
  - ☐ No
- **If you replied No, please specify which questions were unclear and why.**

  ………………………………………………………………………………………………

Appropriateness of Response Options

- **Did the response options provided for each question allow you to express your views accurately?**
  - ☐ Yes
  - ☐ No
- **If No, please suggest any improvements.**

  …………

Impartiality and Lack of Bias

- **Did you find any survey questions or response options leading, biased, or suggesting a particular answer?**
  - ☐ Yes
  - ☐ No
- **If your reply is Yes, please give more details.**

  …………

Acceptability of Response Options

- **Were there any response options that you found confusing, inappropriate, or difficult to choose from?**
  - ☐ Yes
  - ☐ No
- **If your reply is Yes, please give more details.**

  …………

Overall Quality of the Questionnaire

- **Would you consider this questionnaire to be well structured, readable, and relevant?**
  - ☐ Yes
  - ☐ No
- **If your reply is No, please provide any suggestions for improvement.**

  …………

Clarity and Sufficiency of Instructions

- **Were the instructions provided in the questionnaire clear and sufficient to guide you through the survey?**
  - ☐ Yes
  - ☐ No
- **If your reply is No, what additional guidance would have been helpful?**

  …………

Logical Flow of Questions

- **Did the sequence of questions feel logical and easy to follow?**
  - ☐ Yes
  - ☐ No
- **If your reply is No, please specify any abrupt transitions or inconsistencies.**

  …………

Time Required to Complete the Survey

- **Did you feel that the time required to complete the survey was appropriate?**
  - ☐ Yes
  - ☐ No
- **If your reply is No, please specify whether it was too long or too short and suggest an ideal duration.**

  …………

Please share your university email address (*if you agree to attend follow-up interview*): …………

Appendix II: Questionnaire (For Pretesting of Follow-Up Evaluation Survey Instrument)

Introduction

This pretesting questionnaire is designed to evaluate the clarity, logical flow, appropriateness, and other aspects of the follow-up survey before its full-scale administration. The objective is to identify any problems with the survey in the pretesting phase rather than at the actual evaluation phase of the program. The follow-up will be conducted later (after the pretesting) to assess the medium-term impact of the intervention (GenAI integration in academic courses) on participants' knowledge, confidence, application, and awareness. Pretesting helps identify any issues with question wording, response options, or survey structure to ensure a smooth and effective data collection process of the intervention evaluation.

Before proceeding, please read the informed consent statement. Your participation is voluntary, and your responses will remain confidential. Your information like university emails will not be shared with anyone and will be deleted after pretesting report is completed. The final report will be published without identifying details, but with aggregated responses, with select quotes included after removing identifying details *(if quotes are shared by the survey participants)*. Individual responses will be retained for **one month** post-pretesting and then securely deleted. By continuing, you agree to participate in this pretesting phase.

Useful Instructions

The questionnaire is divided into five sections; first four sections represent one of the evaluation aspects: *knowledge, confidence, application, and awareness*. These sections contain the following:

- Evaluation questions designed to measure the medium-term impact of the program on that aspect.

Last section represents the questionnaire-related aspects and contains the following questions:

- Pretesting questions to assess potential issues with the survey instrument, such as clarity, logical flow, appropriateness of response options, impartiality, and time required to complete the survey.

The evaluation questions and pretesting questions include Likert scale items, binary (yes/no) questions, and open-ended questions to gather detailed feedback. If you find any question unclear, difficult to answer, or biased, please provide details in the open-ended sections. Additionally, a follow-up interview may be conducted to further understand your responses and gather any additional feedback you wish to share. If you agree to take part in the follow-up, please mention your university email at the end of the questionnaire; otherwise, leave it blank.

Knowledge

- **How much new have you learned about GenAI since the course?**
  - ☐ 1
  - ☐ 2
  - ☐ 3
  - ☐ 4
  - ☐ 5

Confidence

- **How much confidence have you gained in using GenAI for complex tasks related to work or research?**
  - ☐ 1
  - ☐ 2
  - ☐ 3
  - ☐ 4
  - ☐ 5
- **How much GenAI learning support you have provided to your colleagues, classmates, or friends?**
  - ☐ 1
  - ☐ 2
  - ☐ 3
  - ☐ 4
  - ☐ 5

Application

- **Do you still use GenAI?**
  - ☐ Yes
  - ☐ No

Awareness

- **Have you encouraged others to use GenAI responsibly?**
  - ☐ Yes
  - ☐ No
- **Have you used other GenAI technologies, apart from ChatGPT?**
  - ☐ Yes
  - ☐ No

Questionnaire Aspects for Pretesting

Clarity and Comprehension

- **Did you clearly understand all the questions in this survey?**
  - ☐ Yes
  - ☐ No
- **If you replied No, please specify which questions were unclear and why.**

  ....

Appropriateness of Response Options

- **Did the response options provided for each question allow you to express your views accurately?**
  - ☐ Yes
  - ☐ No
- **If No, please suggest any improvements.**

  ....

Impartiality and Lack of Bias

- **Did you find any survey questions or response options leading, biased, or suggesting a particular answer?**
  - ☐ Yes
  - ☐ No
- **If your reply is Yes, please give more details.**

  ....

Acceptability of Response Options

- **Were there any response options that you found confusing, inappropriate, or difficult to choose from?**
  - ☐ Yes
  - ☐ No
- **If your reply is Yes, please give more details.**

  ....

Overall Quality of the Questionnaire

- **Would you consider this questionnaire to be well structured, readable, and relevant?**
  - ☐ Yes
  - ☐ No
- **If your reply is No, please provide any suggestions for improvement.**

  ....

Clarity and Sufficiency of Instructions

- **Were the instructions provided in the questionnaire clear and sufficient to guide you through the survey?**
  - ☐ Yes
  - ☐ No
- **If your reply is No, what additional guidance would have been helpful?**

  …………

Logical Flow of Questions

- **Did the sequence of questions feel logical and easy to follow?**
  - ☐ Yes
  - ☐ No
- **If your reply is No, please specify any abrupt transitions or inconsistencies.**

  …………

Time Required to Complete the Survey

- **Did you feel that the time required to complete the survey was appropriate?**
  - ☐ Yes
  - ☐ No
- **If your reply is No, please specify whether it was too long or too short and suggest an ideal duration.**

  …………

Please share your university email address (*if you agree to attend follow-up interview):* …………

Appendix III: Questionnaire (Immediate Evaluation Survey Instrument)

Introduction

This questionnaire is designed to evaluate the short-term impact of the intervention (GenAI integration in academic courses) on participants' knowledge, confidence, application, and awareness. This evaluation is conducted with the students immediately after getting exposed to the GenAI in the course.

Before proceeding, please read the informed consent statement. Your participation is voluntary, and your responses will remain confidential. No personal information will be collected, and the evaluation report will be published with aggregated responses of the participants. The final evaluation report will also include the selected quotes shared by the participants after removing identifying details. Individual responses will be retained for **one month** post-completion of immediate evaluation in the university cloud storage and then securely deleted. By continuing, you agree to participate in this evaluation.

Useful Instructions

Please read the instructions carefully before proceeding.

- The questionnaire is divided into four sections; each one represents one of the evaluation aspects: *knowledge, confidence, application, and awareness*. These sections contain the evaluation questions that are designed to measure the short-term impact of the program on that aspect.
- The evaluation questions include 5-point Likert scale items *(1 representing strong disagreement and 5 representing strong agreement)*, binary *(yes/no)* questions, and open-ended questions to gather detailed feedback. You can include detailed insights about your perceptions about the GenAI impact for the evaluation aspect in the open-ended sections.

Knowledge

The knowledge aspect measures your understanding of GenAI concepts, for instance how to write prompts; its capabilities, for instance what it can do for you; and limitations, for instance wrong outputs. The idea here is to see how much basic understanding you have that helps you to understand from a user perspective about what is generative AI, how does it work, how to get the output, and what could be the issues with the technology, at basic interaction levels. It assesses your ability to recognize key terms and fundamental principles.

- **How well do you understand GenAI's capabilities and limitations after completing the course?**
  - ☐ 1
  - ☐ 2
  - ☐ 3
  - ☐ 4
  - ☐ 5
- **What is one key GenAI concept you learned in this course that you did not know before?**

  ……..……

Confidence

This survey measures the impact of the GenAI course on your learning, confidence, and awareness of AI tools immediately after getting exposed to the technology in the course. The confidence aspect measures how comfortable and self-assured you feel when using or explaining GenAI to your peers, at basic levels. This means you may not be an expert but feel some confidence to think about using it or to discuss about this technology. At basic levels, It measures increased self-efficacy in using GenAI for simple tasks and assesses your comfort in trying out GenAI tools.

- **How confident are you in using GenAI for academic work?**
  - ☐ 1
  - ☐ 2
  - ☐ 3
  - ☐ 4
  - ☐ 5
- **How comfortable are you in describing GenAI's use to classmates?**
  - ☐ 1
  - ☐ 2
  - ☐ 3
  - ☐ 4
  - ☐ 5
- **Could you please elaborate on your confidence in using or discussing this technology with your peers, at basic levels?**

  ……………

Application

This survey measures the impact of the GenAI course on your learning, confidence, and awareness of AI tools immediately after getting exposed to the technology in the course. The application aspect measures students' initial use of GenAI in coursework. It assesses whether they have attempted to apply GenAI in assignments or to explore the course topics without using it for final assessments. It also evaluates your ability to integrate basic GenAI outputs into the academic work—seeking further directions, clarity of topics, etc.

- **Did you use GenAI in any coursework or final assessment?**
  - ☐ Yes
  - ☐ No
- **Can you describe a specific task where GenAI was most helpful to you during the course, regardless of whether you used it for the final assessment?**

  ……………

Awareness

This survey measures the impact of the GenAI course on your learning, confidence, and awareness of AI tools immediately after getting exposed to the technology in the course. The awareness aspect measures your basic understanding of ethical AI use and availability of other GenAI technologies.

- **How much basic awareness of ethical AI use you gain during your course?**
  - ☐ 1
  - ☐ 2
  - ☐ 3
  - ☐ 4
  - ☐ 5

- **Can you share any ethical concerns while using GenAI in your course?**

  ……..……

- **Did you become aware of the existence of other GenAI technologies?**
  - ☐ Yes
  - ☐ No
- **Can you share any other GenAI technologies you became aware of during the course?**

  ……..……

Appendix IV: Questionnaire (Follow-Up Evaluation Survey Instrument)

Introduction

This questionnaire is designed to evaluate the medium-term impact of the intervention (GenAI integration in academic courses) on participants' knowledge, confidence, application, and awareness. This evaluation is conducted a few months after students have been exposed to GenAI in the previous course.

Before proceeding, please read the informed consent statement. Your participation is voluntary, and your responses will remain confidential. No personal information will be collected, and the evaluation report will be published with aggregated responses of the participants. The final evaluation report will also include the selected quotes shared by the participants after removing identifying details. Individual responses will be retained for **one month** post-completion of immediate evaluation in the university cloud storage and then securely deleted. By continuing, you agree to participate in this evaluation.

Useful Instructions

Please read the instructions carefully before proceeding.

- The questionnaire is divided into four sections; each one represents one of the evaluation aspects: *knowledge, confidence, application, and awareness*. These sections contain the evaluation questions that are designed to measure the medium-term impact of the program on that aspect.
- The evaluation questions include 5-point Likert scale items *(1 representing strong disagreement and 5 representing strong agreement)*, binary *(yes/no)* questions, and open-ended questions to gather detailed feedback. You can include detailed insights about your perceptions about the GenAI impact for the evaluation aspect in the open-ended sections.

Knowledge

The knowledge aspect measures retained and deepened knowledge about GenAI. It assesses the ability to recall key concepts and in being in processing of more knowledge about the technology beyond what was acquired in the course. It also evaluates progress in grasping GenAI's evolving capabilities and limitations.

- **How much new GenAI-related concepts you have learned through self-study, research, or practical application?**
  - ☐ 1
  - ☐ 2
  - ☐ 3
  - ☐ 4
  - ☐ 5
- **What new concepts about GenAI have you learned beyond what was explored during your course?**

  ....

Confidence

This survey measures the impact of the GenAI course on your learning, confidence, and awareness of AI tools after few months of getting exposed to the technology in the course. The confidence aspect measures your enhanced confidence levels in using GenAI beyond coursework, possibly for more complex tasks, and describing this technology to your peers, for instance making them aware and helping them get a better understanding of the technology through practical use. This measure is not about application, i.e., did you really apply it, but about how confident you feel in doing so.

- **How confident are you in using GenAI for the tasks that required complex problem-solving—where solutions could not be generated solely by technology, but also needed research strategies, critical thinking, and external data?**
  - ☐ 1
  - ☐ 2
  - ☐ 3
  - ☐ 4
  - ☐ 5
- **How much GenAI learning support you have provided to your colleagues, classmates, or friends?**
  - ☐ 1
  - ☐ 2
  - ☐ 3
  - ☐ 4
  - ☐ 5
- **Could you please elaborate on your confidence in using or discussing GenAI with your peers, particularly at a level beyond basic understanding?**

  ....

Application

This survey measures the impact of the GenAI course on your learning, confidence, and awareness of AI tools after few months of getting exposed to the technology in the course. The application aspect measures independent use of GenAI in academic or professional settings. It assesses application in different courses, thesis work, or job-related tasks. It also evaluates the ability to leverage GenAI for complex problem-solving and decision-making.

- **Do you still use GenAI?**
  - ☐ Yes
  - ☐ No
- **Can you describe a specific task (e.g., tasks pertaining to (but not limited to) research, work, other courses, personal projects) where GenAI was particularly helpful to you in the time post-course completion?**

  ……………

Awareness

This survey measures the impact of the GenAI course on your learning, confidence, and awareness of AI tools after few months of getting exposed to the technology in the course. The awareness aspect measures critical awareness of AI, ethics, and other relevant technologies. This means that you should now be fully aware of various ethical aspects, and maybe you have experimented with other GenAI technologies.

- **Have you encouraged others to use GenAI responsibly?**
  - ☐ Yes.
  - ☐ No.
- **Have you used other GenAI technologies, apart from ChatGPT?**
  - ☐ Yes
  - ☐ No
- **Can you share an instance where you or someone you know faced an ethical dilemma related to GenAI?**

  ……………
- **Which other GenAI technologies did you try to experiment with post-course?**

  ……………

MIX
Papier aus verantwortungsvollen Quellen
Paper from responsible sources
FSC® C105338

If you have any concerns about our products,
you can contact us on
ProductSafety@springernature.com

In case Publisher is established outside the EU,
the EU authorized representative is:
Springer Nature Customer Service Center GmbH
Europaplatz 3, 69115 Heidelberg, Germany

Printed by Libri Plureos GmbH
in Hamburg, Germany